Colloid–Polymer Interactions

ACS SYMPOSIUM SERIES **532**

Colloid–Polymer Interactions

Particulate, Amphiphilic, and Biological Surfaces

Paul L. Dubin, EDITOR
Indiana University–Purdue University

Penger Tong, EDITOR
Oklahoma State University

Developed from a symposium sponsored
by the Divisions of Polymer Chemistry, Inc.,
and of Colloid and Surface Chemistry
at the 203rd National Meeting
of the American Chemical Society,
San Francisco, California,
April 5–10, 1992

American Chemical Society, Washington, DC 1993

Library of Congress Cataloging-in-Publication Data

Colloid–polymer interactions: particulate, amphiphilic, and biological
surfaces : developed from a symposium / sponsored by the Division of
Polymer Chemistry, Inc., and by the Division of Colloid and Surface
Chemistry of the American Chemical Society at the 203rd National
Meeting of the American Chemical Society, San Francisco, California,
April 5–10, 1992; Paul L. Dubin, Penger Tong, editors.

    p.   cm.—(ACS symposium series, ISSN 0097–6156; 532)

Includes bibliographical references and indexes.

ISBN 0–8412–2696–2

1. Polymers—Congresses. 2. Colloids—Congresses. 3. Adsorption—
Congresses.

I. Dubin, Paul. II. Tong, Penger, 1956–   . III. American Chemical
Society. Division of Polymer Chemistry. IV. American Chemical
Society. Division of Colloid and Surface Chemistry. V. American
Chemical Society. Meeting (203rd: 1992: San Francisco, Calif.)
VI. Series.

QD380.C62   1993
547.7′04545—dc20   93–8495
                                                                CIP

The paper used in this publication meets the minimum requirements of American National
Standard for Information Sciences—Permanence of Paper for Printed Library Materials, ANSI
Z39.48–1984. ∞

Foreword

THE ACS SYMPOSIUM SERIES was first published in 1974 to provide a mechanism for publishing symposia quickly in book form. The purpose of this series is to publish comprehensive books developed from symposia, which are usually "snapshots in time" of the current research being done on a topic, plus some review material on the topic. For this reason, it is necessary that the papers be published as quickly as possible.

Before a symposium-based book is put under contract, the proposed table of contents is reviewed for appropriateness to the topic and for comprehensiveness of the collection. Some papers are excluded at this point, and others are added to round out the scope of the volume. In addition, a draft of each paper is peer-reviewed prior to final acceptance or rejection. This anonymous review process is supervised by the organizer(s) of the symposium, who become the editor(s) of the book. The authors then revise their papers according to the recommendations of both the reviewers and the editors, prepare camera-ready copy, and submit the final papers to the editors, who check that all necessary revisions have been made.

As a rule, only original research papers and original review papers are included in the volumes. Verbatim reproductions of previously published papers are not accepted.

M. Joan Comstock
Series Editor

Contents

Preface

THE PURPOSE OF THE SYMPOSIUM leading to this book was to bring together a diverse collection of researchers working in fields in which interactions between colloidal surfaces and polymers play a central role. In fact, the symposium participants provided a variety of perspectives, coming from industrial and academic laboratories of chemistry, chemical engineering, materials science, physics, applied biochemistry, and biotechnology, representing research efforts in 10 countries. This broad, international participation indicates the diversity and vitality of activity taking place in the basic and applied aspects of colloid–polymer interactions. Accordingly, this volume reflects the current research trends in this rich field of science.

The chapters in this book deal with specific research situations, from both the experimental and theoretical points of view. To conceptually unify what appears to be a diverse collection of work on disparate systems, an overview chapter by William B. Russel serves as an introduction. This chapter reviews various phenomena such as flocculation, stabilization, phase separation, and rheology of colloidal particles in polymer solutions, which are all recognized as macroscopic manifestations of the short-range forces between polymer segments and colloid surfaces.

This volume consists of four parts. The first part is devoted to theoretical studies and computer simulations. These studies deal with the structure and dynamics of polymers adsorbed at interfaces, equations of state for particles in polymer solutions, interactions in diblock copolymer micelles, and partitioning of biocolloidal particles in biphasic polymer solutions. The second part discusses experimental studies of polymers adsorbed at colloidal surfaces. These studies serve to elucidate the kinetics of polymer adsorption, the hydrodynamic properties of polymer-covered particles, and the configuration of the adsorbed chains. The third part deals with flocculation and stabilization of particles in adsorbing and nonadsorbing polymer solutions. Particular focus is placed on polyelectrolytes in adsorbing solutions, and on nonionic polymers in nonadsorbing solutions. In the final section of the book, the interactions of macromolecules with complex colloidal particles such as micelles, liposomes, and proteins are considered.

The chapters and their bibliographies should provide an up-to-date resource in this interdisciplinary field, encompassing a wide range of technologies. Applications of water-soluble polymers—e.g., in water treatment, paper making, and food additives—involve their interactions with colloid particles more often than not. Rheology control in paints and magnetic media often reflects polymer–particle association. The adsorption of polymers to surfaces is central in oil recovery applications. In biotechnology, polymer–colloid interactions form the basis of novel methods for protein purification and enzyme immobilization. This collection therefore is intended not only for those interested in theory and the behavior of model systems, but also for workers in related technological fields.

Acknowledgments

Acknowledgments are due to several individuals who were involved in the origination and fulfillment of this project. Anna Balazs was co-organizer of this symposium, and Richie Davis, Jim Rathman, Françoise Winnik, and Jiulin Xia were all capable and effective session chairs. Financial support was essential, given the strong participation of overseas scientists, and we are grateful for the assistance of the Petroleum Research Fund; the ACS Divisions of Colloid and Surface Chemistry and Polymer Chemistry, Inc.; Mobay Chemical Company; Johnson's Wax Company; and Aqualon, a Division of Hercules.

PAUL L. DUBIN
Department of Chemistry
Indiana University–Purdue University
Indianapolis, IN 46205–2810

PENGER TONG
Department of Physics
Oklahoma State University
Stillwater, OK 74078

May 23, 1993

Chapter 1

Macroscopic Consequences of Polymer–Particle Interactions

William B. Russel

Department of Chemical Engineering, Princeton University, Princeton, NJ 08544–5263

The following surveys the nature of interactions between soluble polymer and colloidal particles in the context of controlling the bulk behavior of dispersions through polymeric additives. The intent is to relate course-grained macromolecular structure and solution properties to effectiveness in flocculating, stabilizing, or phase separating dispersions and to identify the corresponding effects on the rheology.

Interactions between soluble polymer and either colloidal particles, surfactant micelles, or proteins control the behavior and viability of a large number of chemical and biochemical products and processes. Considerable scientific interest also centers on these interactions because of their profound and, sometimes, unexpected effects on the thermodynamics and dynamics of the dispersions or solutions, known collectively as complex fluids. Syntheses of novel block copolymers, improved scattering and optical techniques for characterization, and predictions emerging from sophisticated statistical mechanical approaches provide additional stimulus. Thus, the area is vigorous academically and industrially as evidenced by the broad and international participation in this volume.

Polymers in these systems function in a variety of ways depending on their molecular structure and concentration, the nature of the solvent, and the characteristics of the other components (1). For homopolymers three primary types of situations exist (Figure 1):

(1) *Adsorption.* Reversible adsorption of individual segments onto a surface perturbs the chain configuration substantially from a random coil. At equilibrium bound and free segments may interchange rapidly within the adsorbed chains, but bound and free chains exchange more slowly. Kinetics can also affect the adsorption step, through slow relaxation of the first chains to arrive from their initially coiled state and slow passage of subsequent chains through the barrier posed by those already adsorbed (2).

(2) *Depletion.* Nonadsorbing chains have little incentive to sacrifice configurational entropy in order to approach a surface closely. Consequently the center of mass remains a radius of gyration or more away from the surface in dilute solutions, leaving a depleted segment density at the surface (3).

0097–6156/93/0532–0001$06.00/0

(3) *Grafting*. In principle and, in some cases, in practice a chain can be tethered to the surface by one end, leaving the rest of the molecule extended toward the bulk fluid. At low graft densities the surface only slightly perturbs the random coil, but at high densities excluded volume interactions cause extension away from the surface toward the pure solvent (2).

The proliferation of copolymers offers the possibility of combining these modes of interaction in several ways (Figure 2):

(1) *Random* copolymers comprised of strongly adsorbing and nonadsorbing groups assume configurations that depend on the detailed structure, e.g. with the size of loops primarily reflecting the distance between adsorbing groups rather than the balance between configurational and enthalpic contributions to the free energy. Most polymeric flocculants and many polymeric dispersants have such structures. If present, surfactants often "decorate" the hydrophobic groups, reducing the tendency to adsorb and perturbing the conformation of the chain in solution.

(2) *Diblock* copolymers act somewhat as polymeric surfactants in selective solvents, forming micelles in solution and adsorbing strongly to surfaces via the less soluble block. The more soluble blocks then behave much as grafted chains. Many dispersants are tailored in this way (4).

(3) Grafting terminal hydrophobes onto water soluble chains or polar groups onto nonpolar backbones produces *triblock* copolymers. In solvents selective for the midblock these associate to form a network in solution, adsorb onto the surfaces of colloidal particles, or absorb into the core of micelles. The reversibility of these complexes varies with size of the end blocks and the selectivity of the solvent and is sensitive to the presence of surfactants. The water soluble versions comprise the associative thickeners cited as a revolutionary advance in the coatings industry (5).

The behavior of these systems hinges on short range intermolecular forces - among polymer segments, solvent molecules, the particle surface, and any surfactant or protein present. In addition, fixed charges on either the polymer backbone or the particle surface introduce strong electrostatic forces, which act over distances varying from microns to nanometers as a function of ionic strength. Thus *nonionic polymers* and *polyelectrolytes* often perform quite differently.

This overview attempts to anticipate some of the subjects addressed in the papers of this volume, providing some background and perspective by drawing heavily on the introductory and concluding portions of Ploehn and Russel (1). Readers interested in more complete treatments should consult that review for theoretical approaches, the treatment of experimental techniques by Cohen Stuart, *et al.* (6), and the elegant presentation of scaling theories by de Gennes (2).

Polymeric Flocculation

Water soluble polymers function as flocculants in water treatment processes, paper making, and minerals processing. The objective is to destabilize colloidal particles, creating large flocs to be removed subsequently by sedimentation, flotation, or filtration. Polymers function without the salt necessary to suppress the electrostatic forces generally controlling colloidal stability in aqueous dispersions. High molecular weights, $10\text{-}10^4$ kg/mole, are advantageous and polyelectrolytes promote adsorption onto particles of opposite charge (7). With nonionic polymers flocculation results from bridging of individual chains across the gap between two particles, producing a deep attractive minimum in the pair potential. This occurs, however, only for partial coverage, implying an optimum at adsorbed amounts of about one half monolayer per surface.

Theoretical analyses of adsorption, assuming that chains equilibrate within the gap but do not exchange with the bulk solution, suggest that low molecular weight polymers adsorb with relatively flat conformations, making bridging across the stabilizing electrical double layers unlikely. For higher molecular weights at moderate

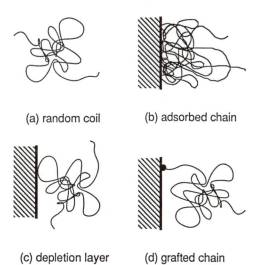

(a) random coil   (b) adsorbed chain

(c) depletion layer   (d) grafted chain

Figure 1. Conformation of homopolymer chains in solution: a, bulk solution; b, adsorbing; c, nonadsorbing; and d, terminally anchored. (Reproduced with permission from reference 1. Copyright 1990 Academic.)

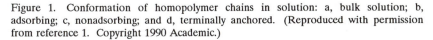

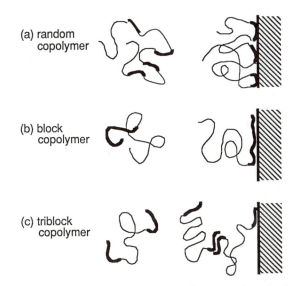

(a) random copolymer

(b) block copolymer

(c) triblock copolymer

Figure 2. Conformation of AB copolymers in solution: a, random copolymers; b, diblock copolymer; and c, triblock copolymer. (Reproduced with permission from reference 1. Copyright 1990 Academic.)

concentrations the depth and range of the attractive minimum in the potential initially increase with surface coverage, but the former decreases to zero at full coverage (8). Hence, bridging flocculation should occur only if the range of the attraction exceeds twice the double layer thickness while the magnitude remains significant, i.e. before restabilization occurs.

Pelssers, *et al.* (9) probed these issues for PEO adsorbed onto polystyrene at different surface coverages. At low molecular weights, the adsorption was allowed to equilibrate before the ionic strength was raised to 10^{-2} M to initiate flocculation. A distinct onset of flocculation appeared at a critical polymer dose, presumably corresponding to the surface coverage at which the range of the attraction due to bridging exceeded twice the double layer thickness. Restabilization was not observed with the limited surface coverage achievable with low molecular weights.

For higher molecular weights, however, the process differs qualitatively with a marked dependence on time, the initial number density of particles, and polymer dosage. Apparently the rate of polymer reconformation during adsorption relative to the rate of particle collisions becomes a key consideration (10,11). Indeed, the data reveals a transition from encounters between particles bearing equilibrium adsorbed layers, at low particle number densities and long collision time scales, to bridging by adsorbed polymer with dimensions comparable to the coil in solution at higher number densities and, hence, shorter collision times. In the latter situation reconformation continues during the flocculation process, causing the rate to decrease monotonically with time and, in some cases, arresting the process with flocs of finite size. The advantages of higher molecular weight polymers must arise in part from such kinetic effects.

Recent simulations by Dickinson and Euston (12) address bridging by very short chains at very low coverages as a function of the strength of attraction. With adsorption energies greater than 1 kT per segment, trains of segments lying on the surface dominate the polymer configurations. Flocs grow as fractals with dimension 1.75 and particles connected by rather short bridges only a few segments long. The shortness of the chains and the weakness of the attraction introduce sufficient reversibility to prevent many flocs from growing very large, so the connection with actual systems is still weak.

For polyelectrolytes, analyses of equilibrium adsorption are only beginning to emerge. Muthukumar's (13) treatment of a single polyelectrolyte chain interacting with a charged surface predicts adsorption with a mean square end-to-end distance of $O(\kappa^{-1})$ below a critical ionic strength. The consequences with respect to the mechanisms just cited and the interaction potentials between two surfaces remain to be explored. Thus, the situation is currently clouded by an incomplete understanding of both the kinetic processes and the equilibrium behavior of polyelectrolytes.

Experimental studies by Peffercorn and coworkers (14) further probe these issues, characterizing the evolving aggregate size distribution via conventional particle counting for polystyrene latices in solutions of partially ionized poly(vinyl pyridine). The results illustrate several transitions, necessitating somewhat speculative interpretations supported by comparison with results for classical fractal aggregation results. At low surface coverages bridging apparently dominates, producing kinetics and aggregate size distributions characteristic of reaction limited aggregation. For slightly higher coverages a similar initial period is followed by an asymptotic regime with a diffusion limited character. The authors attribute this to charge neutralization, through relaxation of the oppositely charged polymer chains into a flat conformation on the surface, permitting dispersion attractions to act as for high ionic strengths in the absence of polymer. For coverages of 0.50 and above flocculation slows considerably and resumes reaction limited behavior, apparently due to bridging controlled by slow reconfiguration of chains between two interacting particles. Clearly the time scales delineating the various regimes must be strong functions of the adsorption energy, polymer molecular weight, particle size, etc.

These studies demonstrate considerable progress in sorting out the relative importance of bridging versus charge neutralization, as affected by the configurational relaxation of the adsorbed chains. One suspects, however, that rational design of processes must still rely on empirical studies of the systems of interest.

Polymeric Stabilization

Interaction potentials calculated from scaling, lattice, and self-consistent field theories for adsorbing homopolymer and terminally anchored layers in good solvents clearly confirm their ability to stabilize colloidal dispersions against flocculation due to dispersion forces. In fact, the practice preceeded the analyses by centuries for adsorbing polymers, with applications dating to ancient times, and decades for block copolymers, which emerged from industrial laboratories in the 1960s (*15*).

Relative to electrostatic stabilization polymeric dispersants offer several advantages:
  (1) no electroviscous effects,
  (2) stability in nonaqueous solvents with low dielectric constants and surface charges,
  (3) robust, long-term stability at high volume fractions, and
  (4) reversible flocculation with changes in solvent quality.
Stability generally requires a good solvent for the stabilizer and a minimum layer thickness to mask the dispersion forces. In addition, adsorbing homopolymers only produce fully repulsive interaction potentials for strong adsorption at full coverage, when the slowness of desorption retains polymer within the gap during interactions between particles. Otherwise, macromolecules squeezed between the surfaces desorb, thereby reducing their free energy and the repulsive potential. Indeed, slow flocculation, or aging, occurs in dispersions sufficiently concentrated to maintain the layers in contact. Thus, homopolymers can be satisfactory but are not the optimum stabilizers.

Chains anchored to the surface, by either chemical grafting or an insoluble block, always produce a repulsion in good solvents. Consequently, copolymers, e.g. diblock, comb, or graft, comprise the most effective stabilizers. Advances in polymer synthesis continue to increase the macromolecules available for this application (e.g.*4*). Direct grafting to the particle is a feasible alternative but requires chemistry specific to the particle.

Two important features of the phenomena addressed by experiment and theory are the layer thickness necessary to provide stability and the conditions at which the dispersions flocculate. The first follows from superposition of the repulsive potential for terminally anchored chains and the attractive dispersion potential, as the layer thickness necessary to maintain the attractive minimum shallower than a few kT. For relatively dense layers of monodisperse chains in good solvents the variation of the repulsive force between flat plates with separation is reasonably well characterized (*16,17*). Conversion via the Derjaguin approximation to spheres produces a potential that dominates the attraction for even weak interactions between the grafted layers; so the minimum occurs at separations of roughly twice the layer thickness. Except for rather low molecular weight stabilizers, the dispersion attraction is retarded by the finite propagation speed of electromagnetic waves and decays as the separation squared. Consequently, the critical layer thickness varies as $[a\lambda A_{eff}/kT]^{1/2}$ with A_{eff} characterizing the magnitude of the dispersion force and λ representing the characteristic wavelength associated with the dielectric relaxations (*18*). Hence, the requisite molecular weight increases with the radius of the particle and the dielectric mismatch between the particle and the solvent, a fact not always appreciated by those choosing dispersants.

A large body of experimental data gathered by Napper and others (*15*) first established a correlation between the condition for incipient flocculation, i.e. the delineation between stable and unstable dispersions, and the theta condition for the soluble block of the stabilizer. Theoretical predictions for the interaction potentials agree for modest surface coverages, e.g. 1-3 monolayers. However, for PDMS chains grafted onto silica particles at surface coverages of 12-18 monolayers Edwards *et al.*. (*19*) observed stability to considerably worse than theta conditions. This phenomena, termed "enhanced steric stabilization" by Napper (*15*), is predicted by mean-field theories as the consequence of three-body correlations among segments at high concentrations (c.f. *18*, Chap. 9). Recent studies are beginning to assess in more detail the effects of adsorbed amount and layer thickness on stability (*20*).

Underlying the application of block copolymers as stabilizers is, of course, the necessity of adsorbing them onto the particles from solution or introducing/growing them during the particle synthesis. The latter serves for the original application, emulsion polymerization of latices. However, success in stabilizing coating formulations or ceramic slips, for example, depends on the former. Here two issues arise:

(i) the kinetics of adsorption from selective solvents (*21,22*).

(ii) the factors controlling the total adsorbed amount (*23,24,25*).

Experiments probing the kinetics for macroscopic surfaces of metal or glass/quartz are diffusion limited at short times, but at longer times reveal extraordinarily complex relaxation processes unless the solution concentration lies below the critical micelle concentration. The adsorbed amounts reported in the latter work exhibit interesting trends with assymmetry of the diblock, in reasonable conformity with expectations from the theory of Marques, et al. (*26*). Thus the basic understanding is advancing quite nicely, but cost and availability still loom as factors limiting applications.

Phase Separations

The alternative to stability is not always flocculation, i.e. a nonequilibrium aggregated state, as implied in the preceeding sections. As with molecular systems, homogeneous dispersions become thermodynamically unstable when an attractive minimum Φ_{min} becomes sufficiently deep to drive the osmotic compressibility to zero. The locus of points on a kT/Φ_{min}-ϕ plot corresponding to this condition comprises the spinodal and lies within the boundary defining a two-phase region. Generally, $\Phi_{min} = 2\text{-}5\,kT$ suffices to cross the phase boundary yet permits dispersions of submicron particles to separate on a reasonable time scale, e.g. in a few days, into two coexisting equilibrium phases. With significantly stronger attractions, equilibration requires up to six months or a year.

Asakura and Oosawa (*3*) first identified depletion as a mechanism for generating an attractive interparticle potential. Numerous elaborations of their simple model followed, including sophisticated lattice and self-consistent field theories. Recently, Evans (*27*) resolved some inconsistencies in the evaluation of the effective pair potential between the original niave model and the subsequent detailed analyses and achieved quantitative consistency between predictions and the detailed experiments employing bilayer membranes in a micropipette device.

Phase separations induced by dissolved nonadsorbing polymer were recognized and exploited in the 1930s to concentrate, or cream, rubber latices (*16*, §15.2) and reappeared in the 1970s in the formulation of coatings (e.g.*28*). Experiments with a variety of dispersions - polymer latices, silica spheres, and microemulsions - in aqueous or nonaqueous polymer solutions at either ideal of non-ideal conditions now clearly establish the coexistence of equilibrium phases, one dense and one dilute in particles, above a critical polymer concentration corresponding to an attractive minimum of $O(2\text{-}3\,kT)$. As expected from simple theory, the critical polymer concentration decreases with increasing molecular weight or particle size (e.g. *29,30,31*).

Subsequent studies revealed two modes of phase behavior depending on the ratio of the particle radius a to the polymer's radius of gyration r_g (*32,33,34*). For example, electrostatically stabilized polystyrene latices in aqueous solutions of dextran exhibit a simple fluid - solid transition for $a/r_g = 6.9$, but a more complex phase diagram with a fluid - fluid envelope, critical and triple points, and a fluid - solid region for $a/r_g = 1.9$ (*33*). The solid phases display iridescence, characteristic of crystalline order, and the expected mechanical responses, i.e. yield stresses and finite low-frequency elasticity; the fluid phases are opaque and flow as Newtonian liquids in the low shear limit (*35*).

At modest concentations of free polymer, all dispersions behave generally as described above, independent of the mechanism stabilizing the particles in the absence of polymer, e.g. grafted polymer chains or electrostatic repulsion. However, with increasing polymer concentration, the correlation length and, hence, the depletion layer thickness decrease, so the range of the repulsion ultimately becomes important.

For particles stabilized with grafted chains, free polymer penetrates further into the grafted layer with increasing concentration, causing Φ_{min} to pass through a maximum and then decrease (*36*). Thus the phase or stability boundary becomes an envelope, with destabilization at low polymer concentrations and restabilization at higher values. Edwards, *et al.* (*19*) dispersed silica spheres bearing a layer of grafted polystyrene chains in a solution of somewhat higher molecular weight polystyrene in toluene and observed aggregation over a finite range of polymer concentrations for volume fractions of particles exceeding a critical value, roughly as expected from the theory.

Thus nonadsorbing polymer can induce phase separations in colloidal systems, with the nature of the phases depending primarily on the ratio of the particle and polymer sizes. Since the strength of the attraction is not necessarily a monotonic function of the polymer concentration, e.g. because of penetration of the free polymer into a grafted layer, both destabilization and restabilization are possible.

A second means of suppressing this phase transition was developed for water-based coatings formulations, in which polymer latices comprise the film-forming component and soluble polymer serves as a thickener. Sperry, *et al.* (*5*) recognized that grafting terminal hydrophobes onto the soluble chains would generate reversible associations, in the form of micelles, within the solution itself and reversible adsorption onto the hydrophobic surfaces of the polymer latices. The adsorption reduces or eliminates the depletion layer but also presents the possibility of bridging flocculation. The resulting phase diagram, generated for a model paint, identifies a small bridging envelope at low polymer and surfactant concentrations and the conventional depletion phase separation at surfactant concentrations sufficient to suppress adsorption.

Recently Santore, *et al.* (*37,38*) quantified Sperry's conceptual model via a statistical mechanical theory, first deriving an effective pair potential from analyses of the configurations of linear chains capped with stickers between spheres and then calculating the thermodynamic properties and phase boundaries via a standard perturbation theory. With full equilibrium between chains in the gap and those in the bulk solution, moderate adsorption of the terminal groups can produce a purely repulsive potential, differing markedly from the attractions always expected for adsorbing or nonadsorbing homopolymers. With stronger segment-surface attractions an attractive minimum appears at separations on the order of r_g , but then disappears with increasing bulk concentration due to excluded volume interactions between adsorbed layers.

This combination of features seen for nonadsorbing polymer, adsorbing homopolymer at full equilibrium, and grafted polymers yields interesting phase behavior as suggested by Sperry, *et al.* (*5*) and delineated in Figure 3. Note the single phase fluid at low polymer concentrations, a two-phase envelope due to bridging at intermediate values, and the return to a single phase fluid at higher polymer concentrations. The latter region is the desirable one from the standpoint of coatings formulations. Then the associative polymers are available to affect rheological properties as desired, without inducing phase separation by either bridging or depletion.

Rheology

Soluble polymers have long been exploited to adjust the rheology of colloidal dispersions and new possibilities based on novel macromolecular structures continue to emerge. Here we review selectively the effects of grafted polymer, adsorbing homopolymer, associative polymers, and nonadsorbing polymer.

The primary effect of a grafted layer of thickness L is to increase the effective hydrodynamic and thermodynamic sizes of a colloidal particle, assuming good solvent conditions. Consequently, dispersions of these particles behave as non-Newtonian fluids with low and high shear limiting relative viscosities (η_o and η_∞), and a dimensionless critical stress ($a^3\sigma_c/kT$) that depend on the effective volume fraction. As for hard spheres the viscosities diverge at volume fractions ϕ_{mo} and $\phi_{m\infty}$, respectively, with $\phi_{mo} < \phi_{m\infty}$; for $\phi > \phi_{mo}$, the dispersions yield and flow as pseudoplastic solids. However, ϕ_{mo} and $\phi_{m\infty}$ vary with L/a, the softness of the repulsion.

The data of Mewis et al. (39) and d'Haene (40) for poly(methyl methacrylate) spheres stabilized by poly(12-hydroxy stearic acid) and dispersed in decalin correlate reasonably well with results for hard spheres for low to moderate volume fractions, although the critical stress is somewhat smaller. For highly concentrated dispersions, however, packing constraints cause some interpenetration of the layers at rest and viscous forces at high shear rates drive the particles even closer together. Consequently, the effective layer thickness decreases with increasing ϕ and Pe, the dimensionless shear rate.

The grafted layer also affects two other features of the rheology. First, thicker polymer layers enhance the elasticity due to the longer range of the repulsion relative to the hard core size. Thus, samples formulated at $\phi > \phi_{mo}$ possess easily measurable static elastic moduli and offer a rheological means of characterizing the interparticle potential. Second, the softer repulsion apparently suppresses the shear thickening transition encountered at high volume fractions for the harder particles.

With adsorbing homopolymer the phenomena differ significantly since the polymer can either stabilize or bridge colloidal particles. In addition, the viscous forces generated by shear can either induce bridging or breakup flocs formed by bridging. Otsubo and co-workers (41-45) illustrate the possibilities with silica spheres in solutions of poly(acrylamide) in glycerin - water mixtures over a range of particle sizes and volume fractions, polymer molecular weights and concentrations, solvent quality, and shear rates.

In glycerin, the dispersions are stable and behave as a shear thinning fluids. However, the viscosity initially decreases with the addition of silica, only increasing at higher concentrations. Adsorption reduces the solution viscosity by the depletion of polymer, but enhances the relative viscosity by increasing the hydrodynamic volume of the particles. Since the particles adsorb ~1.5 mg/m^2 of polymer on a surface area of 0.13 m^2/mg, 5 wt% silica lowers the solution concentration by 1 wt%. At higher silica concentrations, interactions among the coated particles increase the viscosity, suggesting a layer thickness of ~10nm, and lead to shear thinning at higher shear rates. Thus, the coated particles behave as stable spheres at all shear rates.

Better solvents, e.g. glycerin - water mixtures, enhance the possibility of bridging and render the rheological behavior quite sensitive to those parameters controlling the surface coverage: the molecular weight and concentration of the polymer, the size and concentration of the particles, and the solvent quality. At high surface coverages, achieved with large particles, low silica concentrations, or high polymer molecular weights, the dispersions exhibit stability and relatively low viscosities. But rather low coverages, arising for smaller particles, higher silica concentrations, or lower molecular weights, permit bridging and produce highly pseudoplastic materials. Most interesting then are intermediate coverages, for which an irreversible shear-induced transition occurs, transforming a low viscosity dispersion into a paste at a critical stress.

Subsequent analyses support this interpretation and correlate the elasticity of the samples with bridging through simulations and analogies to percolation processes (*46*).

With nonadsorbing polymer, significant rheological transitions correlate with the phase boundaries mentioned above (*35*). The slowness of the macroscopic phase separation permits rheological characterization of a metastable structure that changes little over time for samples formulated within the two-phase region. The systems respond pseudoplastically, but the microstructure recovers to a reproducible rest state after shear.

The initial steady shear measurements identified several features of the phenomena:

(1) an apparent yield stress appeared at the fluid-solid phase boundary and increased with distance into the two-phase region and

(2) discontinuous shear thinning appeared near or within a fluid - fluid envelope, though the low shear viscosity remained Newtonian.

However, rheometry on such dispersions is fraught with difficulties. For example, Buscall, *et al.* (*47*) subsequently reported much lower and essentially Newtonian viscosities at low shear rates for formulations exhibiting a fluid-solid transition. When challenged that slip at the wall might be the culprit, they machined Couette devices with much larger gaps and repeated one set of measurements. The results reveal a much higher but still Newtonian low shear viscosity, suggesting that both previous sets of measurements were confounded by a stress-controlled slip process!

As a final example, let's consider the rheological effects of the associative polymers formulated in the single phase fluid region at high concentrations (*5*). The solutions themselves respond viscoelastically as classical reversible networks (*48*) well characterized by a single relaxation time, associated with the dynamics of junction formation and breakup. The steady shear viscosities exhibit a Newtonian low shear plateau, a modest degree of shear thickening (for relatively low molecular weights), and pronounced shear thinning. The shear thickening clearly derives from extension of the chain beyond the linear elastic limit, but either nonaffine deformation of the network or rupture of the junctions by viscous stresses could produce the shear thinning.

In the commercial formulations the presence of particles at $\phi=0.20$ merely increases the level of the viscosity and shifts the characteristic shear rates and frequencies. Jenkins' data (*48*) conforms at $\phi < 0.10$, but at higher volume fractions the Newtonian plateau disappears. This raises the question of whether a reversible polymer network persists or the particles simply interact as fuzzy and, perhaps, sticky spheres with much slower dynamics that control the shear rate. Additional data exists beyond that cited here but does not seem to resolve this issue.

Similar possibilities have been demonstrated for nonpolar continuous phases (*49,50*). With AOT w/o microemulsions having decane as the continuous phase, poly(isoprene) with terminal poly(ethylene oxide) blocks associates with the water droplets. Under conditions that promote bridging between droplets the microemulsions become orders of magnitude more viscous and appear to undergo a sol-gel transition (Figure 4). This presents an ideal system for detailed study since conductivity, dynamic light scattering, and dielectric measurements, among others, readily probe the dynamics and microstructure.

Thus soluble polymer, interacting in a controlled fashion with colloidal particles, can transform both the equilibrium state and the mechanical properties of dispersions. The possibilities range from equilibrium, low viscosity fluids to nonequilibrium, pseudoplastic pastes with high yield stresses. However, substantial gaps still exist in the ability to, for example, (i) create high viscosity equilibrium fluids with prescribed relaxation spectra, (ii) impart a sol-gel transition at prescribed conditions, or (iii) connect explicitly macromolecular structure with rheological behavior.

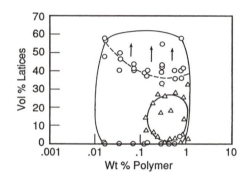

Figure 3. Phase behavior of 0.24 μm (○) and 0.068 μm (△) acrylic latices containing poly(ethylene oxide) of M_n = 35–40 kg/mole with terminal hydrophobes. (Reproduced with permission from reference 38. Copyright 1990 Royal Society of Chemistry.)

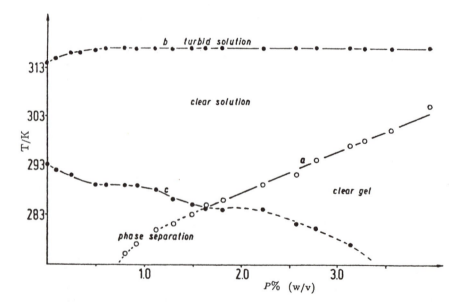

Figure 4. Phase diagram of AOT microemulsion containing a poly(ethylene oxide)/poly(isoprene)/poly(ethylene oxide) copolymer illustrating the polymer induced phase separation and gelation boundaries. (Reproduced with permission from reference 49. Copyright 1989 Elsevier.)

Literature Cited

1. Ploehn, H.J., and Russel, W.B. *Adv. Chem. Eng.* **1990** 15, 137.
2. de Gennes, P.G. *Adv. Colloid Interface Sci.* **1987** 27, 189.
3. Asakura, S., and Oosawa, F. *J. Chem. Phys.* **1958** 22, 1255;*J. Polym. Sci.* **1954** 33, 183 .
4. Reiss, G., Bahadur, P., and Hurtvez, G. *Encycl. Polym. Sci. Eng.* **1987** 2, 324.
5. Sperry, P.R., Thibeault, J.C., and Kostanek, E.C. *Adv. Coatings Sci . Technol.* **1987** 9, 1.
6. Cohen Stuart, M.A., Cosgrove, T., and Vincent, B. *Adv. Colloid Interface Sci.* **1986** 24, 143.
7. Rose, G.R., and St. John, M.R. *Encycl. Polym. Sci.* **1985** 7, 211.
8. Scheutjens, J.M.H.M., Fleer, G.J., and Cohen Stuart, M.A. *Colloids Surf.* **1986** 21, 285.
9. Pelssers, E., Cohen Stuart, M.A., and Fleer, G.J. *Colloids Surf.* **1989** 38, 15.
10. Gregory, J. *Colloids Surf.* **1987** 31, 231.
11. Dickinson, E., and Eriksson, L. *Adv. Colloid Interface Sci.* **1991** 34, 1.
12. Dickinson, E., and Euston, S.R. *J. Chem. Soc. Faraday Trans.* **1991** 87, 2193.
13. Muthukumar, M. *J. Chem. Phys.* **1987** 86, 7230.
14. Elaissari, A., and Pefferkorn, E. *J. Colloid Interface Sci.* **1991** 141, 522.
15. Napper, D.H. "Polymeric Stabilization of Colloidal Dispersions." Academic Press, New York, 1983.
16. Patel S.S., and Tirrell, M. *Ann. Rev. Phys. Chem.* **1989** 40, 597.
17. Taunton, H.J., Toprakcioglu, C., Fetters, L.J., and Klein, J. *Nature (London)* **1988** 332, 712.
18. Russel, W.B., Saville, D.A., and Schowalter, W.R. "Colloidal Dispersions." Cambridge Univ. Press, London, 1989.
19. Edwards, J., Lenon, S., Toussaint, A.F., and Vincent, B. *ACS Symp. Ser.* **1984** 240, 281.
20. Wu, D.T., Yokoyama, A., and Setterquist, R.L. *Polym. J.* **1991** 23, 709.
21. Munch, M.R., and Gast, A.P. *J. Chem. Soc. Far. Trans. I* **1990** 86, 1341; *Macromolecules* **1990** 23, 2313.
22. Leermakers, F.A.M., and Gast A.P. *Macromolecules* **1991** 24, 718.
23. Gast, A.P. *Proc. NATO ASI Ser., Ser. E* **1990** Strasbourg, France.
24. Parsonage, E., Tirrell, M., Watanabe, H., and Nuzzo, R.G. *Macromolecules* **1991** 24, 1987.
25. Hair, M.L., Guzonas, D., and Boils, D. *Macromolecules* **1991** 24, 341.
26. Marques, C., Joanny, J.F., and Leibler, L. *Macromolecules* **1988** 21, 1051.
27. Evans, E.A. *Macromolecules* **1989** 22, 2277.
28. Sperry, P.R., Hopfenberg, H.B., and Thomas, N.L. *J. Colloid Interface Sci.* **1981** 82, 62.
29. Cowell, C., Li-In-On, R., and Vincent, B. *J.C.S. Faraday I* **1978** 74, 337.
30. Vincent, B., Luckham, P.F., and Waite, F.A. *J. Colloid Interface Sci.* **1980** 73, 508.
31. de Hek, H., and Vrij. A. *J. Colloid Interface Sci.* **1981** 84, 409.
32. Sperry, P.R. *J. Colloid Interface Sci.* **1984** 99, 97.
33. Patel, P.D., and Russel, W.B. *J. Colloid Interface Sci.* **1989a** 131, 192.
34. Vincent, B. *Chem. Eng. Sci.* **1987** 42, 779.
35. Patel, P.D., and Russel, W.B. *J. Colloid Interface Sci.* **1989b** 131, 201.
36. Gast, A.P., and Leibler, L. *Macromolecules* **1986** 19, 686.
37. Santore, M.M., Russel, W.B., and Prud'homme, R.K. *Macromoleucles* **1990** 23, 3821.
38. Santore, M.M., Russel, W.B., and Prud'homme, R.K. *Faraday Discuss. Chem. Soc.* **1990** 90, 323.

39. Mewis, J., Frith, W., Strivens, T.A., and Russel, W.B. *AIChE J.* **1989** 35, 415.
40. D'Haene, P. PhD Dissertation, Katholieke Universiteit Leuven, Belgium, 1992.
41. Otsubo, Y., and Umeya, K. *J. Colloid Interface Sci.* **1983** 95, 279.
42. Otsubo, Y., and Umeya, K. *J. Rheol.* **1984** 28, 95.
43. Otsubo, Y. *J. Colloid Interface Sci.* **1986** 112, 380.
44. Otsubo, Y., and Watanabe, K. *J. Non-Newtonian Fluid Mech.* **1987** 24, 265.
45. Otsubo, Y., and Watanabe, K. *J. Colloid Interface Sci.* **1988** 122, 346.
46. Otsubo, Y., and Nakane, Y. *Langmuir* **1991** 7, 1118.
47. Buscall, R., McGowan, I.J., and Mumme-Young, C.A. *Faraday Discuss. Chem. Soc.* **1990** 90, 115.
48. Jenkins, R.D., Silebi, C.A., and El-Aasser, J.S. *Proc. ACS Div. Polym. Matls.* **1989** 61, 629.
49. Eicke, H.F., Quellet, C., and Xu, G. *Colloids Surf.* **1989** 36, 97.
50. Quellet, C., Eicke, H.F., Xu, G., and Hauger, Y. *Macromolecules* **1990** 23, 3347.

RECEIVED May 25, 1993

THEORY AND SIMULATION

Chapter 2

Kinetics of Polymer Adsorption and Desorption
Poly(ethylene oxide) on Silica

J. C. Dijt, M. A. Cohen Stuart, and G. J. Fleer

Department of Physical and Colloid Chemistry, Agricultural University
Wageningen, Dreijenplein 6, 6703 HB Wageningen, Netherlands

Expressions are developed for the rate of transfer of polymer molecules towards and from an adsorbing surface under the condition of a fully relaxed adsorbed layer. The special form of the adsorption isotherm for polymers leads to rather simple expressions for the rate of adsorption (up to ~ 85% saturation) and for the rate of desorption (in the neighbourhood of saturation). We compare these results with experimental data that we obtained using novel techniques (reflectometry combined with impinging jet flow for adsorption and streaming potential in a capillary for desorption,respectively) .Good overall agreement is found, which implies that PEO layers on silica equilibrate rapidly.

Linear polymers are typically very flexible objects that may undergo major changes in shape upon adsorption. For the kinetics of the adsorption process this would seem to imply that reconformation is an important step, and therefore one expects to see effects characteristic for transport in dense polymer systems: slow relaxation due to topological constraints, pinning, reptation etc. Of course, such processes can only be studied experimentally if they are rate-determining, i.e. *slower* than the rate at which molecules are transported towards or from the surface.

It is of course also conceivable that the surface processes are *faster* than the mass transport. In that case we cannot observe any of the surface processes, since the surface is fully equilibrated with the adjacent solution, and all that remains is a concentration gradient in the solution which controls the kinetics. In order to be able to interpret experimental data we must therefore have a prediction of the adsorption and desorption rates for the equilibrated case. Deviations from this are then indicative of slow surface processes.

However, most explicit kinetic equations proposed in the literature are based on rather simple adsorption models which have no relevance for polymer adsorption. In this paper we will therefore first consider mass transfer-limited polymer adsorption and desorption rates from a theoretical point of view. We will then turn our attention to measurements which were designed in such a way as to enable accurate control over the mass transfer rate, so that data can be meaningfully analyzed. We used two different methods. The first is reflectometry combined with impinging-jet flow in order to measure adsorbed mass as a function of time.This

0097–6156/93/0532–0014$06.00/0

method works well for adsorption. Rates of polymer desorption, however are usually extremely small so that they cannot ,be picked up by reflectometry. We therefore developed streaming potential measurements from which we get the time dependence of the *thickness* of the adsorbed layer. This thickness appears to be extremely sensitive to subtle changes in adsorbed mass (*1*).

Theory
Provided hydrodynamic conditions are stationary and well-defined, the steady-state mass transport through a solution towards the adsorbing surface can always be written as

$$J = k (c_b - c_s)   \tag{1}$$

where k is a (known) function of the hydrodynamic variables (flow rate, geometry of the flow cell) and polymer properties (diffusion coefficient). Furthermore, c_b is the bulk concentration and c_s is the polymer concentration in the solution immediately adjacent to the wall.Explicit expressions for k can be found in the literature (*2*) For the particular case of a liquid jet emerging from a cylindrical channel in a flat plate, and impinging on a second plate parallel to the first one, Dabros and Van de Ven (*3*) derived for the stagnation point

$$k = 0.776 (\alpha \bar{v})^{1/3} (D/R)^{2/3}   \tag{2}$$

where $\bar{v}$ is the mean velocity in the inlet tube, R the radius of the tube, and D the polymer's diffusion coefficient. The flow intensity parameter α can be calculated numerically, once the diameter of the inlet tube, the distance between the parallel plates, the flow rate, and the kinematic viscosity of the liquid are known.A similar expression is available for flow through a cylindrical capillary. In this case k does not only depend on hydrodynamic parameters and on D, but also on the lateral position x with respect to the inlet of the capillary:

$$k = 0.855 (\bar{v}/xR)^{1/3} D^{2/3}   \tag{3}$$

where R is again the capillary radius; the equation is due to Lévêque (*4*).

Since we shall be dealing with the case of a completely relaxed surface layer, we can make use of the fact that this layer is in full equilibrium with the adjacent solution , i.e. with c_s. Hence, the relation between the adsorbed amount Γ and c_s at any moment is nothing else than the equilibrium adsorption isotherm. Of course the equilibrium is *local* in the sense that the surface is not equilibrated with the entire solution, but only with the narrow zone in its immediate vicinity. A typical polymer adsorption isotherm (here obtained from the Scheutjens -Fleer theory (*5*)) is shown in Figure 1.It has the familiar shape: very steep initial rise ('high affinty ') followed by a nearly horizontal pseudo-plateau. The consequences of this particular isotherm shape for the rates of adsorption and desorption are at least qualitatively easily seen by combining Figure1 . with eq.1 : the concentration difference $c_b - c_s$ is indicated in the figure by means of an arrow. For the case of adsorption the arrow has a length corresponding to just c_b up to something like 85% of the final adsorption. Hence, we expect $J = d\Gamma/dt$ to be constant over this range and then to drop quickly as Γ comes very close to saturation: integrating eq. 1 with $c_s = 0$ simply gives $\Gamma = kc_bt$. The experimental check would thus consist of measuring $\Gamma(t)$ which should be linear up to almost saturation, with a slope given by kc_b, i.e. determined by hydrodynamic parameters and the polymer molecular weight (diffusion coefficient).

For desorption the picture is quite different. Starting from the saturated layer, c_S falls rapidly from c_b to very low values during the first few percents of desorption. Since c_b is zero for the usual desorption experiment this implies that after an intial small desorption, the rate of transport becomes extremely small. In order to carry the analysis somewhat further we need to know more about the isotherm shape at the very low concentrations hidden under the rising part of the curve in Figure 1. Theoretical work (6) has led to the rather general conclusion that for interacting homopolymers on a surface the isotherm takes the form:

$$\Gamma = \Gamma_0 + p \log (c_S / c_{S,0})$$ (4)

where p is a small coefficient ($\approx$ a few percent per decade). Inserting this into eq. 1 and integrating we obtain

$$\Gamma = \Gamma_0 - p \log (t / \tau)$$ (5)

where $\tau = 0.43 \ p/kc_{S,0}$ is a characteristic time, typically well below 1 sec; we may take $c_{S,0}$ and Γ_0 as the concentration and adsorbed amount, respectively, at the start of the experiment. Equation 5 has interesting consequences. First of all it becomes immediately clear that polymer desorption is an inherently slow process; this comes from the logarithmic time dependence. Also, our derivation shows that desorption of polymers is not difficult because of kinetic barriers at the surface, but simply due to the extremely high affinity nature of the isotherm, and the concomitently low driving force for diffusion away from the surface. The higher the molecular weight of the polymer, the lower the value of p and the slower the desorption proceeds. Finally, we note that eq. 4, describing the static behaviour of an adsorbed polymer, is strikingly similar to eq. 5, which describes the dynamics; the logarithmic terms only differ in their signs. This suggests an interesting experimental check: one should not only look at the behaviour of $\Gamma(t)$ on its own, but also compare it with $\Gamma(c_S)$: symmetry between the two would be evidence that the adsorbed layer is truly in equilibrium on the time scale of the experiment.

Materials and Methods
Adsorbed amounts were determined by means of reflectometry using a simple set-up (7). Light from a small (0.5 mW) polarized HeNe laser hits a piece of an (oxidized) silicon wafer at the stagnation point, under the Brewster angle for Si/H_2O (70°).The reflected beam is separated into its two polarization components by means of a polarizing beam splitter and the intensities of both components are detected by photodiodes.The ratio of the two intensities gives, after proper calibration, the total mass of adsorbed polymer per unit area. An impinging jet flow is achieved by injecting solvents or solutions through a cylindrical channel in a prism which has its base parallel to the reflecting substrate.
        For desorption experiments, we measured streaming potentials in the following way. (The entire experiment is described more extensively elsewhere (8)). A glass capillary (length 20 cm, inner diameter 0.30 mm) is mounted between a flask at the outletside and to an electronic valve at the inlet side. The latter is connected to two bottles containing solvent and polymer solution, respectively. The bottles are pressurized with N_2. Via the valve either of the two solutions is injected into the capillary.The flow is laminar for nearly the whole length of the capillary.and the wall shear rate is 11300 s^{-1}.The streaming potential is measured via 2 reversible Ag/AgCl-electrodes, one at either end of the capillary. The

hydrodynamic layer thickness δ_h is calculated from the streaming potential V_s using the following equation

$$\delta_h = \kappa^{-1} \ln (V_{s,0}/V_s) \qquad (6)$$

where κ^{-1} is the Debye length (10 nm in this solvent, 1.00 mM NaCl), and $V_{s,0}$ the streaming potential of the bare surface, which varied between 140-155 mV.

Monodisperse poly(ethylene oxide) purchased from Polymer Laboratories (see table 1) was dissolved without further purification and the solution was stirred for one night at room temperature before use.

Table 1. Characteristics of PEO samples

Code	M$\underline{w}$, kg/mol	M$\underline{w}$/M$\underline{n}$
7	7.1	1.03
23	23	1.08
56	56	1.05
105	105	1.06
400	400	1.08
847	847	1.16

Results and Discussion

By way of example we present in fig 2 an adsorption curve as obtained from the reflectometry experiment.The vertical scale gives the signal S with respect to the (pure solvent) baseline; S is simply proportional to Γ.It is evident from this figure that for the narrow fraction of PEO employed here, the curve has indeed a linear part which extends up to ~ 85% of the adsorbed mass at saturation. In the figure we also indicate the point where injection of solution was stopped and pure solvent was injected.Within the limit of detection, we cannot observe any desorption in this experiment. However, if we carry out a similar experiment with low MW PEO (eg., 10^4), we do observe a (slow) desorption. This demonstrates the lower affinity, and therefore higher p-value, of the lower molecular weights .

The influence of the variables determining k can now be checked by measuring the initial slopes under different conditions. The effects of concentration and flow rate can be seen in fig 3, where (calibrated) initial slopes of a set of experiments are plotted logarithmically against the logarithm of the Reynolds number Re, which is a measure of the flow rate. The experimental results are indicated by crosses, and the adsorption rates given by eqs.1-2 are represented by continuous curves. As is clear from this figure, the dependences, both on c_b and on $(\alpha \bar{v})^{1/3}$,predicted by eq. 2 agree very well with the data. The absolute rate seems to be slightly lower than expected, but this may be due to averaging over a finite zone around the stagnation point, which tends to lower the detected adsorbed mass, or to small errors in the determinations of geometric parameters determinig the flow pattern. We are therefore strongly inclined to conclude that the measured adsorption rate is entirely determined by the rate of mass transfer.

This conclusion is further corroborated by data presented in Figure 4, where the effect of the molecular weight is shown. As expected, $d\Gamma/dt$ decreases with increasing M, since the larger molecules diffuse more slowly. More quantitatively, we expect the diffusion coefficient to vary as the inverse of the hydrodynamic radius R_h. For long chains in a good solvent R_h is known to scale as $M^{3/5}$, for

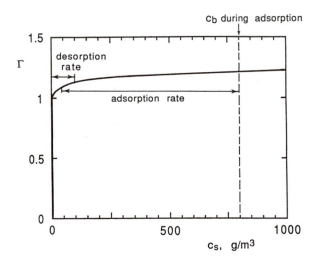

Figure 1. Theoretically calculated adsorption isotherm for a polymer of 100 segments adsorbing from a theta solvent ($\chi=0.5$). The isotherm was obtained from the self-consistent field theory of Scheutjens and Fleer. Two arrows indicate the values of the concentration difference c_b-c_s between bulk and surface zone: one for adsorption (high c_b) and one for desorption (c_b =0). The length of the arrows is a measure of the rate of the corresponding processes.

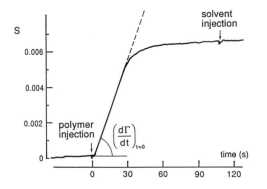

Figure 2. Typical reflectometric adsorption curve. Start of polymer and solvent injection, respectively, are indicated by arrows. The initial slope is given by kc_b with k given by eq. 2. $M = 400$ kg/mol, $c = 10$ mg/dm^3, $Re = 12.2$ and $dS/d\Gamma = 8.9 \times 10^{-3}$ m^2/mg

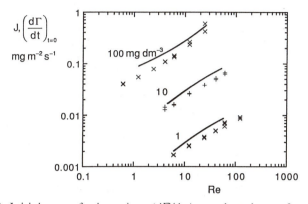

Figure 3. Initial rate of adsorption, $(d\Gamma/dt)_{t=0}$, plotted as a function of the Reynolds number Re (double logarithmic plot), for 3 different polymer concentrations indicated in the figure. Crosses: experimental data. Solid curves: kc_b with k calculated from eq. 2.

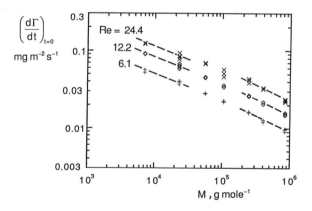

Figure 4. Initial rates of adsorption plotted as a function of molecular weight (double logarithmic plot), for 3 different Reynolds numbers as indicated in the figure. Dashed lines were drawn according to scaling laws: -1/3 at low M and -2/5 at high M.

shorter, ideal chains we should have $M^{1/2}$. Combining this with the dependence of J on $D^{2/3}$ we expect to get $M^{-1/3}$ at the low M end of the range and $M^{-2/5}$ at high M, and a smooth crossover in between (9).The data in Figure 4 agree strikingly well with these scaling laws. Indeed, it seems that we are observing a transport rate-limited process.

We now turn to the desorption experiments. Figure 5 gives a typical result obtained from the streaming potential measurement; measured potentials were converted into hydrodynamic layer thicknesses using eq.6. Before starting the experiment the capillary was first flushed with solvent. At t=0 polymer injection started and δ_h increased. After some time (60 sec. in this example) the capillary surface became saturated with polymer and the thickness reached a stable plateau,the height of which is very sensitive to the concentration of polymer in the solution.It is seen in Figure 5 that the rising part of the curve is not at all linear, as in the reflectometry experiment, but has a kind of S-shape. There are two reasons for this. Firstly, $d\delta_h/d\Gamma$ is not constant but increases strongly with Γ, especially near saturation. Secondly, the rate of polymer deposition on the capillary wall is a function of the distance x from the capillary entrance (eq. 3); as a consequence the entrance side of the capillary is always closer to a state of saturation than the exit side, so that the layer thickness, and the corresponding drop in the streaming potential are initially dominated by the entrance side. With help of model calculations we could however make sure that also in this experiment, the increase of δ_h with time agreed fully with the transport equation.

As soon as solvent is injected, δ_h drops sharply due to desorption of the polymer. The rate of this decrease is initially rapid, as can be seen, but soon takes a much lower pace. Upon reinjection of polymer, the initially measured layer thickness was rapidly restored.The first point to note is that in contrast to Γ, δ_h is indeed remarkably sensitive to exposure to pure solvent : very small desorbed amounts give rise to considerable decreases in layer thickness. Another interesting aspect is that the experiment is entirely reversible: adsorption/desorption cycles can be repeated many times without observable differences.

In Figure 6b we report results from a set of static experiments. These were carried out by succesively injecting polymer solutions of decreasing concentration, and measuring the final thickness that was reached at each concentration. These results constitute what may be called a thickness isotherm, and are plotted against log c_b , as suggested by eq. 4. The corresponding dynamic experiment consisted simply in recording δ_h as a function of (log) time during continued injection of solvent ; these results are presented in fig 6a.(The initial sections of these curves were dashed because they deviate due to dead-volume effects in the experiment). It is immediately clear that the two sets of curves display remarkable symmetry: they are mirror images of each other (due to the difference in sign only) but otherwise completely similar. The slopes of the curves are given by p $d\delta_h/d\Gamma$, from which we can make an estimate of p values. We find quite reasonable values, comparable to those predicted by theory. As expected, p is smaller for the higher molecular weights; that these samples nevertheless display larger thicknesses is due to the dramatic increase in $d\delta_h/d\Gamma$ at high M. Again, this result is consistent with eqs. 4 and 5, which corroborates the mass transfer-limited character of the desorption process.

Conclusions
The rates of adsorption and of desorption of polyethylene oxide from aqueous solution onto glass were measured with help of new experimental techniques

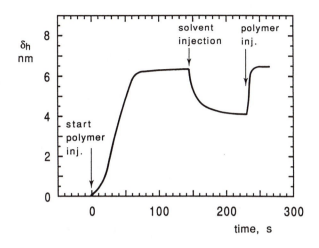

Figure 5. Typical example of streaming potential experiment, showing dependence of layer thickness δ_h on time during initial deposition (rising part), saturation (horizontal part), followed by desorption (decreasing part) and readsorption (rising part). M = 105 kg/mol, c = 10 g/m³ and Re = 127.

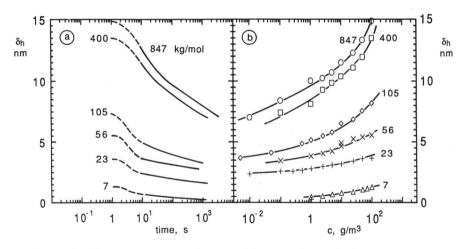

Figure 6. a.Time dependence of polymer layer thickness δ_h (logarithmic time scale) during desorption into pure solvent for 6 different PEO samples. Molecular weights are indicated in the figure. At t = 0 injection of pure solvent on a polymer layer initially saturated at c = 100 g/m³ starts. Initial parts of these curves are dashed since they are affected by instrumental artifacts (dead volume).

b. Concentration dependence of δ_h for the same samples as in a. Note the symmetry between the kinetic data (a) and the static data (b).

which were designed in such a way as to have accurate control over mass transfer rates in solution. For adsorption measurements, reflectometry, combined with impinging jet flow proved to be very suitable. Desorption could be followed with unprecedented precision with help of streaming potential measurements on single glass capillaries; these measurements give the hydrodynamic thickness of the adsorbed polymer layer. We find that all the experimental results could be entirely explained from the rates of mass transfer in the solution, under the assumption that the adsorbed layer is always fully equilibrated with the solution in its immediate vicinty.

REFERENCES

1. Cohen Stuart, M.A.; Waajen,F.H.W.H.; Cosgrove,T.; Vincent,B.; Crowley,T.L. *Macromolecules* **1984**, *17,* 1825.
2. Adamczyk,Z.; Dabros,T.; Czarnecki,J.; Van de Ven,T. *Adv. Colloid Interface Sci.* **1983** , *19*, 183.
3. Dabros,J.; Van de Ven,T. *Colloid Polymer Sci.* **1983**, *261*, 694.
4. Lévêque,A. *Ann. Mines* **1928**, *13*, 284.
5. Scheutjens,J.M.H.M.; Fleer,G.J. *J.Phys.Chem.* **1979,** *83,* 1619.
6. Van der Linden,C.C.;Leermakers,F.A.M. *Macromolecules* **1992,** *in press.*
7. Dijt,J.C.;Cohen Stuart,M.A.; Fleer,G.J. *Colloids Surfaces* **1990,** *51,* 141.
8. Dijt,J.C.; Cohen Stuart,M.A.; Fleer,G.J. *Macromolecules* **1992,** *in press.*
9. Nyström,B.; Roots,J. *Prog. Pol. Sci.* **1982,** *8,* 333.

RECEIVED March 24, 1993

Chapter 3

Adsorption from a Mixture of Short and Long Chains

M. Daoud and E. Leclerc

Laboratoire Léon Brillouin, Commissariat a L'Energie Atomique–Centre National de la Recherche Scientifique, C.E.N. Saclay, 91191 Gif-sur-Yvette Cédex, France

We consider the adsorption of long chains, with N monomers, disolved in a matrix of shorter chains, with P units. We assume that the interaction d of a monomer and the surface is small. For p=1, one has the usual adsorption problem. For PN, one is dealing with a melt, and there is no adsorption. Therefore, changing P allows to monitor the adsorption. We will discuss the various regime that may be found when one changes the length of the short chains. We also consider the normal extension of the long chains and their eventual localization close to the adsorbing surface. For higher concentrations, the existence of trains and large loops is discussed following a recent theory by Joanny and Marques.

Polymers at interfaces have been under continuous study for some time because of both their practical importance and their fundamental properties (1). In the following, we will be interested in adsorbed chains on flat surfaces. Interest in this question arose some time ago, and was renewed recently. Theoretically, the introduction of scaling methods by de Gennes (2), Pincus (3), Binder, Kremer and Eisenriegler (4) allowed a new approach to the problem. Experimentally, new methods such as neutron reflectivity (5) and small angle neutron scattering (6) provided a much more detailed information than was previously available. The combination of both improved tremendously our understanding of the adsorption of linear chains on a flat impenetrable surface when the surface attraction is not too large. Many important questions however remain open. These are related either to the quality of the surface, or to the nature of the solvent. Most actual surfaces are neither completely flat nor homogeneous. Work is in progress to take into account surface roughness (7,8) and chemical imperfections. (9,10) Similarly, the quality of the solvent was considered in many directions. The latter include surface driven phase separation (11) as well as the effect of a mixed solvent. Another solvent effect that will be considered here is related to the length of the solvent molecules. Previous studies were concerned with solutions in a simple solvent (12). It is possible both theoretically and experimentally to consider long polymers, made of N monomers, dissolved in shorter chains, with P units, of similar nature and slightly different interaction with the surface. The interest in this question is twofold. First by changing P, it is possible to have either adsorbed or desorbed probe chains as we shall see. Second, a generalization of this problem is to consider a mixture of both long and

0097–6156/93/0532–0023$06.00/0

short chains in a simple solvent. The question then is the displacement of an adsorbed polymer by another one. In what follows, we will only consider the equilibrium properties for the first case, in the absence of any simple solvent. As we shall see, this question is already very rich.

We will first recall the various bulk regimes for long probe chains dissolved in shorter polymers. Section 3 will review the case of adsorbed chains in the presence of a molecular solvent. Adsorption from a dilute solution in a matrix of short chains will be considered section 4. The last section will deal with the possibility that the large adsorbed chains may be localized near the surface and the eventual presence of large loops in a semi-dilute solution.

2. Long polymers in a matrix made of shorter chains.

In this section, we consider first the structure of a long chain, made of N monomers, dissolved in a matrix of shorter polymers of a similar species but with P units. P may be varied from 1 to N. When P is of order unity, we assume the monomer to be a good solvent of the probe chain. Therefore the latter is swollen. In the other limit, when $P \approx N$, the system is similar to a melt, and the long chain behaves as a Gaussian. It is therefore interesting to consider the cross-over that occurs when the length P of the matrix is varied. This was considered by de Gennes (13) some years ago in a Flory approximation (14,15). The free energy of the long chain may be written in the following way:

$$F = \frac{R^2}{R_0^2} + \frac{v}{P}\frac{N^2}{R^3} \tag{1}$$

where
$$R_0 \sim N^{1/2} \tag{2}$$

and R are respectively the ideal and actual radii of the probe chain, and the excluded volume parameter v is assumed to be positive. The first term in the free energy is the elastic contribution. The presence of P in the interaction energy is related to the Edwards screening (16,17). Minimizing (1) with respect to R we find

$$R \sim N^{3/5} P^{-1/5} \tag{3}$$

Comparing relations (2) and (3), we find the cross-over value P* for the length of the short chains between the swollen and ideal behaviors:

$$P^* \sim N^{1/2} \tag{4}$$

Therefore when P is smaller than the above value, the probe chains are swollen, whereas for larger values, the excluded volume interactions are screened and the long chain is ideal.

It is possible to make one step further in the analysis in the swollen case. This is done by evaluating the interaction term F_{int} in the free energy (1). Assuming the chain is ideal, we find

$$F_{int} \sim v N^{1/2} P^{-1} \tag{5}$$

When the interaction energy is smaller than unity, it may be neglected, and the chain is gaussian. On the other hand, when it is large, one has to take it into account, and to minimize the free energy, leading to relation (3). This provides the cross-over relation (4) for P. It is possible to do one step further when P<<P*, and to define ideal blobs. This is done by considering portions of the long chain made of n segments. Generalizing relation (5), the interaction for such portions is

$$F_{int}(n) \sim vn^{1/2} P^{-1} \qquad (6)$$

Even though the interaction energy of the whole chain is large for P<<P*, it is possible to find a value g_{id} for the portions such that $F_{int}(g_{id}) \sim 1$. This defines the ideal blobs: the interaction inside a blob is not sufficiently large, so that the chain is locally ideal. Let χ be the radius of the ideal blob. We have

$$\chi \sim g_{id}^{1/2} \sim P \qquad (7a)$$

and

$$R \sim \{\frac{N}{g_{id}}\}^{3/5} \chi \qquad (7b)$$

Combining (7a) and (7b), one recovers relation (3). Note that for $P \sim P*$, the blob has the same size as the chain.

If we now consider a finite fraction C of monomers belonging to large chains, it is possible to determine a phase diagram that summarizes the various possible regimes. A C* line separates the dilute regime where the above discussion is basically valid from a semi dilute region where screening of the excluded volume effects takes place because the large chain overlap. Extrapolating relation (3), one finds

$$C* \sim N^{-4/5} P^{3/5} \qquad (8)$$

For higher concentrations, the large chains are characterized by two lengths, namely a screening length ξ and its radius R. Both may be evaluated by scaling considerations. We assume

$$R \sim N^{3/5} P^{-1/5} f_{R,\xi}(C/C*) \qquad (9)$$

where the functions $f_{R,\xi}$ behave as power laws in the semi-dilute range. The exponents of these power laws are determined by conditions that ξ is independent of N, and that R has a gaussian variation. We find

$$\xi \sim C^{-3/4} P^{1/4} \qquad (10)$$

and

$$R \sim N^{1/2} C^{-1/8} P^{-1/8} \qquad (11)$$

For $P \sim C^{-1}$, ξ becomes of the same order as χ, and a second cross-over occurs to a regime for high concentrations, where the probe chains are ideal for all length scales. Note that although the chain is ideal, there is still a screening length

$$\xi \sim C^{-1} \tag{12}$$

in this regime. It corresponds to the distance between contacts of long chains.

3. Adsorption in the presence of a simple solvent.

Before we consider the adsorption of the long probe chain we studied in last section, we review the results that were obtained by scaling arguments (3,12) in the case of a simple solvent. Thus we consider a dilute polymer solution in a good solvent. We assume that the surface is impenetrable and slightly attractive to the monomers: there is a free energy gain δ per monomer on the surface. We assume the latter to be smaller than unity. Let us consider first the case of a single chain. When it is not adsorbed, the number N_S of monomers on the surface is

$$N_S \sim N^{3/5} \tag{13}$$

where the exponent was estimated by computer simulations (17-19) and renormalization group calculations. Therefore the free energy gain F_a per chain is

$$F_a \sim \delta N^{3/5} \tag{14}$$

This is to be compared with an entropy loss of order unity, corresponding to the fact that the center of mass of the polymer becomes localized in the vicinity of the surface. Comparing the former two energies, we find a cross-over from a free to an adsorbed state for the chain for

$$\delta^* \sim N^{-3/5} \tag{15}$$

Thus, although the energy gain per monomer is small, the total energy gain per chain may be very large, and the chain adsorbs. The configuration of the polymer may be obtained by a simple scaling argument. Assuming

$$R \sim N^{3/5} g(\delta N^{3/5}) \tag{16}$$

the characteristic sizes of the chain $R_{/\!/}$ and $R_\perp$ respectively along and orthogonal to the surface are determined by the condition that the adsorbed chain is quasi two dimensional, and has a pancake shape along the surface. Assuming that the functions $g_{/\!/,\perp}$ behave as power laws, we find

and
$$R_{/\!/} \sim N^{3/4} \delta^{-1/4} \qquad\qquad (\delta N^{3/5} \gg 1) \tag{17a}$$

$$R_\perp \sim \delta^{-1} \qquad\qquad (\delta N^{3/5} \gg 1) \tag{17b}$$

Using the same kind of arguments as in section 2, it is possible to define isotropic blobs (20), that are parts of the chain that are on the verge of being adsorbed. These are made of g_{iso} monomers and have a size D such that

$$g_{iso} \sim \delta^{-5/3} \tag{18a}$$

and

$$D \sim g_{iso}^{3/5} \sim R_{\perp} \tag{18b}$$

The chain may be considered as a two dimensional array of isotropic blobs:

$$R_{/\!/} \sim \{\frac{N}{g_{iso}}\}^{3/4} D \tag{19}$$

Using relations (18) and (19) one easily recovers relation (17a).

When the bulk concentration C_b is non vanishing, several surface regime may be found, including a surface semi-dilute regime, where loops larger than D develop because of excluded volume effects, and a plateau regime where the surface is saturated with isotropic blobs. Because of the excluded volume effects, large loops are present, and the chains extend to a distance on the order of the radius of gyration R of a free chain. The surface concentration ϕ_S is calculated knowing that isotropic blobs saturate the surface.

$$\phi \sim \frac{g_S}{D^2} \tag{20}$$

where g_S is the number of monomers per isotropic blob on the surface:

$$g_S \sim g^{3/5} \sim \delta^{-1} \tag{21}$$

Combining equations (20), (21) and (17b), we find

$$\phi_S \sim \delta \tag{22}$$

The concentration profile was calculated by de Gennes and Pincus (3) and by Eisenriegler et al. (4), who found three regimes as a function of the distance z to the surface.

For small distances, in the proximal regime, $l \ll z \ll D$

$$\phi(z) \sim \phi_S z^{-1/3} \tag{23a}$$

For intermediate distances, in the central region, $D \ll z \ll R$

$$\phi(z) \sim z^{-4/3} \tag{23b}$$

This is the most interesting regime because it is the only one present when δ is of order unity, and also because it is self-similar: In this region, the concentration is in the semi-dilute regime. Therefore there is a screening length $\xi(z)$ that is a function of the local concentration $\phi(z)$.

$$\xi(z) \sim \phi(z)^{-3/4} \tag{23c}$$

Because there is only one length scale in the problem, the latter has to be identified with the distance z to the surface. Identifying (23c) with z leads directly to the concentration profile, relation (23b).

Finally, in the distal region, at large distances, for z>>R, the concentration profile falls off exponentially.

Note that all these regimes occur even when the bulk concentration C_b is in the dilute range. When C_b is in the semi-dilute regime, the above discussion is valid for distances smaller than the bulk correlation length . For larger distances, the concentration is the average bulk concentration. There is still however an open question related to the normal extension of the polymers: are they localized in the vicinity of the surface, within a distance ξ, or do they extend to distances of the same order as their radius in the bulk? The answer to this question is related to the formation of large loops in the adsorbed layer and will be discussed in section 5.

4. Adsorption of long chains from a shorter matrix.

Before we come to the discussion of loops, we would like to look at the adsorption of long chains dissolved in a matrix of shorter polymers of similar nature (21), such as discussed in section 2. We assume that short and long chains are not identical, and that a monomer from a long chain on the surface has a free energy gain δ. Because the latter is small, we do not expect any adsorption when P is large. The system then is similar to a polymer melt. The various polymers are equivalent, and no chain may have a finite fraction of its monomers on the surface. When P is of order unity, on the other hand, the system is identical to the one discussed in the previous section. Therefore, it is possible by changing P to cross-over from an adsorbed to a desorbed state for the probe chain.

4a. The single probe.
Let us consider the various states of one probe chain diluted in shorter matrix. For simplicity, we will assume that it is grafted by one endpoint to the surface. We consider first the case when P is small, and the probe is swollen. It is possible to evaluate the number N_s of monomers on the surface: Every ideal blob has $g_{id}^{1/2}$ monomers on the surface. The number of blobs in contact with the wall is $(N/g_{id})^{3/5}$. Therefore

$$N_s \sim \{ \frac{N}{g_{id}} \}^{3/5} g_{id}^{1/2} \tag{24}$$

and the free energy gain F_s is

$$F_s \sim N_s \, \delta \tag{25}$$

This is compared with that of a solvent chain

$$F_p \sim P \tag{26}$$

Comparing (25) and (26) provides us the scaling parameter

$$\Delta = \frac{F_s}{F_p} \sim \delta N^{3/5} P^{-6/5} \tag{27}$$

When Δ is smaller than unity, the long chain does not adsorb. In the opposite limit, on the contrary, it adopts a pancake shape along the surface. It is worth discussing briefly our central condition (27). This is obtained by comparing the energy gain of the probe chain with the energy PkT of a solvent chain in the bulk. The reason for doing this is to have a constant reference energy, corresponding to monomeric solvent. Let us stress here that this is a conjectural assumption. A different possible assumption would be to compare F_S with unity, the latter corresponding to the entropy loss of the center of mass of the probe chain when it is adsorbed. Clearly, this would change the scaling parameter Δ and lead to results different of those that will be discussed below. Such assumption however would imply that the same conditions are valid to adsorb a chain from a dilute solution in a theta solvent and from a melt. This seems to be questionable to us. Therefore, although there might still be discussion about relation (27), we will accept it in the remaining part of the paper. The conformation of the adsorbed chain may be discussed along the same lines as in section 3: The number N_S of units directly on the surface may be calculated with a scaling assumption. Indeed, we may write this number in the following scaled form:

$$N_S \sim N^{3/5} P^{-1/5} f(\delta N^{3/5} P^{-6/5}) \tag{28a}$$

In the adsorbed state, this number is proportional to N. Assuming that in this limit, $f(x)$ behaves as a power law, we find

$$N_S \sim N \delta^{2/3} P^{-1/3} \qquad (\Delta \gg 1) \tag{28b}$$

When D is larger than unity, as mentioned above, the probe chain adopts a pancake shape along the surface for large distance scales. For shorter distances however, it remains isotropic. On a local scale, inside the isotropic blob discussed above, relation (7), it is still ideal. For intermediate distances, the interaction with the surface is not sufficiently strong to adsorb the chain, and we may define adsorption blobs, as in relation (18). These are defined by generalizing equation (27). This leads to the number g_{iso} of elements in the adsorption blobs.

$$g_{iso} \sim P^2 \delta^{-5/3} \tag{29}$$

The size D of the latter blobs is

$$D \sim \{\frac{g_{iso}}{g_{id}}\}^{3/5} \chi \sim P \delta^{-1} \tag{30}$$

The long chain itself may be considered as a two dimensional array of the latter blobs.

$$R_{/\!/} \sim \{\frac{N}{g_{iso}}\}^{3/4} D \sim N^{3/4} \delta^{-1/4} P^{-1/2} \tag{31}$$

It is also interesting to note that the "local" surface concentration ϕ_S of monomers from the probe chain depends on the area where it is defined. In the ideal blob, we have :

$$\phi_S \sim \frac{gid^{1/2}}{\chi^2} \sim P^{-1} \tag{32a}$$

whereas for distances equal to or larger than D, it is

$$\Phi_S \sim \phi_S \{\frac{giso}{gid}\}^{3/5} \{\frac{D}{\chi}\}^{-2} \sim \delta P^{-1} \tag{32b}$$

Finally, we note that desorption of the probe chain occurs when $R_{\parallel} \sim \chi$, when $\Delta \sim 1$. When this is realized, the ideal blob is still much smaller than both lengths.

4b. The plateau regime.

We consider now the case when the surface is saturated with isotropic blobs from long chains. As discussed previously, this may be realized even when the bulk concentration in probe chains is very small. In the plateau regime, the surface is covered with isotropic blobs. Therefore the surface concentration is Φ_S. Note that this is averaged over a distance D in the longitudinal directions. Averaging over a smaller surface area would lead to a smaller result because the ideal blobs are not dense in the surface. The concentration profile is readily calculated.

Inside the ideal blob, the concentration is constant : $\phi = \Phi_S$
For distances within the proximal range, we have

$$\phi(z) \sim \Phi_S \{\frac{z}{\chi}\}^{-1/3} \qquad (\chi << z << D) \tag{33}$$

in the central range, the profile is still self-similar. The profile is evaluated in a similar way as in section (3). We find

$$\phi(z) \sim z^{-4/3} P^{1/3} \qquad (D << z << R) \tag{34}$$

For large distances, in the distal range, the profile falls off exponentially.

These laws may be checked either by neutron small angle scattering on polymers adsorbed on a porous medium or by neutron or X ray reflectometry. The number N_S of monomers directly on the surface may also be measured by N.M.R. or by infra-red spectroscopy.

5. Loop formation.

Last section dealt with the concentration profile of the adsorbed large polymers. We saw that in a dilute bulk solution and in the plateau regime, it extends to distances of the order of the radius of a chain. In a semi-dilute solution, we expect the same profile to be present up to distances of the order of the screening length ξ_b. The former approach only tells us that the blobs are adsorbed as long as

$$d >> d^* \sim gc^{-3/5} P^{6/5} >> 1 \tag{35}$$

where we have generalized equation (15) to the bulk concentration blob with size x, relation(10) and number of elements g_c such that

$$x \sim g_c^{3/5} P^{-1/5} \tag{36}$$

Note that condition (35) implies that for sufficiently large concentrations, the blobs desorb. Using equations (10) and (35), we find the cross-over concentration $\tilde{C}$ for desorption of blobs

$$\tilde{C} \sim d^{4/3} P^{-1} \tag{37}$$

Therefore, for $C > \tilde{C}$ the chains are desorbed, and they extend to a distance of the order of their free radius R.

For concentrations below $\tilde{C}$ and for distances larger than ξ, the former approach does not tell us whether or not the adsorbed macromolecules extend beyond a distance ξ. The first approach to this question was to assume that for $C>>C^*$, the solution is similar to a melt, if the (concentration) blob with size ξ is taken as a unit. If such is the case, we know that because δ is small, there is no adsorption of the chain. Therefore the extension normal to the surface is given by equation (11) for bulk concentration above C^*. This may be questioned however because for intermediate bulk concentrations, $C^*<< C << \tilde{C}$ the blobs are strongly adsorbed, so that the chain of blobs is in strong adsorption conditions. Another related question that was raised by Marques and Joanny (22) is to know the extension of the adsorbed polymers. Indeed, it is possible that in a restricted concentration range above C^*, the chains might be localized within a distance ξ from the surface, and that large loops start developing only for concentrations higher than some value, to be determined. In what follows, we follow their approach to loop formation. Let us first consider concentrations slightly larger than C^*. It is possible to show that within a distance ξ_b there is more than one chain. We consider the case $\delta \approx 1$ for simplicity, Then the surface concentration is, from relation (32b)

$$\Phi_s \sim P^{-1} \tag{32c}$$

The contribution of one blob may be evaluated. The number g_s of monomers on a surface ξ^2 is

$$g_s \sim \frac{g_c}{g_{id}} g_{id}^{1/2} \sim C^{-5/4} P^{-1/4} \tag{33}$$

Therefore the number n_s of chains in a surface ξ^2 is

$$n_s \sim \Phi_s \frac{\xi^2}{g_s} \sim C^{-1/4} P^{-1/4} \tag{34}$$

Because of the self-similarity of the profile, only one of these chains may escape from the surface layer of width ξ. The probability of escape of a chain is therefore proportional to n_s^{-1}. Let a *train* be the part of successive monomers of a chain that are located within the surface layer. Its average length t is

$$t \sim n_s g_s \sim C^{-3/2} P^{-1/2} \tag{35}$$

The total number of trains is equal to the number L of loops. Because we are considering a semi-dilute solution, we know that we have

$$L \sim \{\frac{N}{gc}\}^{1/2} \tag{36}$$

Therefore, the total number T of monomers in the trains is

$$T \sim Lt \sim N^{1/2} C^{-7/8} P^{-7/8} \tag{37}$$

This is to be compared with the maximum number, N/P, of monomers of the probe on the surface. (The presence of P in the latter relation is due to the fact that only $P^{1/2}$ monomers of an ideal blob may be on the wall). Therefore below a concentration C2 the total number of monomers in the trains is independent on C, and is proportional to N. Above C2, this does not hold anymore, and N_0 is proportional to $N^{1/2}$, as in a melt. Comparing T with N/P, we find

$$C_2 \sim N^{-4/7} P^{1/7} \tag{38}$$

It is possible to show, following Marques and Joanny, that if one considers the adsorbance Γ, that is the total number of monomers per unit surface linked to the surface, this may be split into two parts: the first one, Γ_S, comes from the surface layer. The second one comes from outer distances.

$$\Gamma_S = \int_0^\infty \{\phi(z) - \phi(\infty)\} \, dz \tag{39}$$

The adsorbance may be approximated by

$$\Gamma = \Gamma_S + CR_N \tag{40}$$

For $P \ll N^{1/2}$, $P \ll C^{-1}$, and $\delta \approx 1$, we find

$$\Gamma = a^{-2} \{ 1 + N^{1/2} C^{7/8} P^{-1/8} \} \tag{41}$$

The latter relation shows that for $C \ll C_2$, most of the contribution to the adsorbance comes from the surface layer. For higher concentrations, this is inverted. We conclude by noting that except for some rare large loops extending to the radius R of the probe chains, for $C \ll C_2$ most of the large chain is localized in the surface layer, within a distance ξ. This implies that mostly loops smaller than ξ are present. Note also that the total number of monomers $N_0 \sim N\Phi_S/\Gamma$ in direct contact with the surface is proportional to N in this regime.

$$N_0 \sim N/P \qquad\qquad (C \ll C_2) \tag{42a}$$

$$N_0 \sim N^{1/2} C^{-7/8} P^{-7/8} \qquad\qquad (C_2 \ll C \ll P^{-1}) \tag{42b}$$

Finally, we note that for $C \sim P^{-1}$ relation (42b) crosses over to the classical melt value $N_0 \sim N^{1/2}$, and that for $C \sim \tilde{C}$ for $N \sim P^2$.

For $C > C_2$ large loops are present. For $C \ll C_2$, the number of loops (and trains) decreases, and becomes on the order of unity for a concentration C_1. Therefore, below C_1, the chain is made of one single train. The latter is made of all the monomers. Equating the length t of a train, relation (35), to N, we find

$$C_1 \sim N^{-2/3} P^{1/3} \tag{43}$$

We summarize the various regimes. For $C^* \ll C \ll C_1$, the chain is made of one single train localized within a distance ξ from the surface. There might be a finite number of loops larger than ξ present, but they contribute a vanishingly small contribution to the adsorbance: the adsorbed mass is located within ξ. For $C_1 \ll C \ll C_2$, loops larger than ξ develop, but the mass is still located within ξ. The total number of monomers on the surface is still proportional to N. For $C \gg C_2$, large loops are present and extend to the radius of the free chains. A finite fraction of the adsorbed layer is in the diffuse layer between ξ and R. Note that the concentration blobs are adsorbed for $C_2 \ll C \ll \tilde{C}$, and desorbed for $C > \tilde{C}$. It not clear however that there is such a discontinuity in the localization of the monomers, and that the distance where a finite fraction is localized jumps for $C \sim C_2$ from ξ to R. An alternative solution would be that the localization width Λ increases continuously from ξ to R as the concentration is varied from C_1 to C_2. Using as a scaling assumption

$$\Lambda = \xi\, f(C/C_1) \tag{44}$$

Assuming that f(x) behaves as a power law, and that Λ is of the order of R, relation (11), for $C \sim C$, and using relations (38) and (43) we get (23)

$$\Lambda = N\, C^{3/4} P^{-1/4} \tag{45}$$

for the localization length, a surprising result because of the proportionality of the length to N. The existence of such length still has to be shown experimentally. This may be done for instance by quenching the monomers on the surface and diluting the bulk solution with a simple good solvent.

6. Conclusion.

We discussed the adsorption of long chains dissolved in a matrix of shorter and similar polymers. We assumed that the energy gain δ per monomer of the long chain on the surface is small. We found that several regimes are present. In the dilute bulk range, the most interesting regime is the plateau, where the surface is saturated. The concentration profile includes, in addition to the proximal and central zones that are present in a simple solvent, an ideal region in the immediate vicinity of the surface, with width dependent on the mass of the matrix. In the semi dilute bulk regime, we found that the probe chain is localized in the vicinity of the surface for concentrations smaller than C_1. For concentrations between C_1 and C_2, the number of trains increases. Thus large loops are present, although most of the mass of the adsorbed chains remains within the adsorption layer of width D. There might be an extra length Λ in this region, corresponding to the size of the large loops. Such length is proportional to ξ at C_1 and to the radius of the probe chains at C_2. We find $\Lambda \sim NC^{3/4}P^{-1/4}$, much smaller than the radius of the long chain, but proportional to N. For concentrations larger than C_2, the number of monomers on the surface is proportional to $N^{1/2}$, rather than to N, as it was below C_2. Large loops extending to the radius of the chain are present and contain a finite fraction of the adsorbed mass. Finally, above $\tilde{C}$, the blobs are no longer adsorbed, and the chain is as in a melt. Two remarks may be made at this level: for adhesion purposes, the regime between C_1 and C_2 corresponds to a case where the probe chain has N monomers on the surface and therefore is well anchored on the wall. On the other hand it has few large loops. These are present above the concentration C_2, but then the chain has only $N^{1/2}$ monomers on the surface. Thus one is led to make a balance between these contradictory properties, namely a large extension or a very good adhesion.

This bimodal system is interesting because it also allows the study of the displacement of polymer by another one, a work that is under current study. In order to do this one has to introduce a simple solvent in addition to the two species that we considered here.

References.

1. Adsorption from solutions, Academic Press, (1983), R.H. Ottewill, C.H. Rochester, A.L. Smith, eds.
2. P.G. de Gennes, J. Physique, **37**, 1445, (1976).
3. P. Pincus, P.G. de Gennes, J. Physique Lett., **44**, 241, (1983).
4. E. Eisenriegler, K. Binder, K. Kremer, J. Chem. Phys., **77**, 6296, (1982).
5. L. Auvray, J.P. Cotton, Macromolecules, **20**, 202, (1987).
6. E. Bouchaud, B. Farnoux, X. Sun, M. Daoud, G. Jannink, Europhys. Lett., **2**, 315, (1986).
7. D. Hone, H. Ji, P. Pincus, Macromolecules, **20**, 2543, (1987).
8. F. Brochard, J. Physique, **46**, 2117, (1985).
9. S.F. Edwards, Y. Chen, J. Phys. **A21**, 2963, (1988).
10. C. Marques, Thesis, Université Louis Pasteur, (1989).
11. S. Leibler, Thesis, Université Paris XI, (1984).
12. E. Bouchaud, Thesis, Université Paris XI, (1988).
13. P.G. de Gennes, J. Pol. Sci. , Pol. symp. **61** , 313 , (1977) .
14. P.J. Flory, Principles of polymer chemistry . Cornell University press , (1953) .
15. P.G. de Gennes,Scaling concepts in polymer physics , Cornell University press, (1979)
16. S.F. Edwards, Proc. Phys. Soc. **88** , 265 , (1966) .
17. K. Binder, K. Kremer, in Scaling Phenomena in Disordered Systems, NATO A.S.I. B 133, 525, R. Pynn and A. Skjeltorp eds.,Plenum Press (1985).
18. T. Ishinabe, J. Chem. Phys., **76**, 5589, (1982), and **77**, 3171, (1983).
19. K. Kremer, J. Phys. **A16**, 4333, (1983).
20. E. Bouchaud, M. Daoud, J. Physique, **48**, 1991, (1987).
21. M. Daoud, Macromolecules, **24**, 6748, (1991).
22. C. M. Marques, J. F. Joanny, J. Physique, **49**, 1103, (1988). See also C. M. Marques, Thesis, Université Louis Pasteur, Strasbourg, (1989).
23. M. Daoud, G. Jannink, J. Physique, J. Physique II, **1**, 1483, (1991).

RECEIVED January 14, 1993

Chapter 4

Equation of State for Nanometric Dispersions of Particles and Polymers

O. Spalla and B. Cabane

Equipe Mixte Commissariat a L'Energie Atomique–R.P., C.E.N. Saclay, Service de Chimie Moléculaire Bât 125, 91191 Gif sur Yvette Cédex, France

Equilibrium phase diagrams for aqueous dispersions containing particles and adsorbing polymers are reported. The components are nanometric ceria particles and polyacrylamide macromolecules. The phase diagram consists of a 2 phases region surrounded by a one phase region. At low concentrations the samples in the one phase region are sols ; at high concentration they are gels. The equation of state for sols and gels has been determined through osmotic compression.

Mineral particles may be dispersed in water through repulsive forces which keep the particles apart from each other. The most common forces are electrostatic interactions which arise from ionization of the particles surface. They are opposed by Van der Waal attractions which tend to pull particles together. The most important parameter in this balance of repulsions and attractions is the heigth of the potential barrier of repulsion which results from electrostatic interaction; classical DLVO theory indicates that long term stability is ensured if this potential is at least 15 kT (1).

Dispersions of nanometric particles are unusal because the potential barriers are low, since each particle carries a small number of effective charges. Moreover the range of repulsions is often rather short, because the dispersions have a very large surface area and this leads to high concentrations of ions in the aqueous medium. A consequence of these 2 features is that control of nanometric dispersions through electrostatic interactions is notoriously difficult.

An alternative to electrostatic control is the use of adsorbing macromolecules. It is well known that particle surfaces which are saturated with polymer strands tend to repel each other, whereas starved surfaces attract each other because they gain by sharing polymer strands (2).

With nanometric particles the problem of polymer adsorption is different from the usual adsorption on macroscopic surfaces (Spalla O, Cabane B. *Coll. Pol. Sci.*, in press). For one thing, nanometric dispersions have an extremely large surface area, hence they can adsorb a substantial amount of polymer before being saturated. In fact the amount of bound polymer is often comparable to the amount of particles in the system ; thus the system may be better described as a two-component system, similar in nature to mixed solutions of two polymers which interact with each other. This paper presents an analysis of such mixed particles+polymer dispersions from the point of view of two-component systems.

0097–6156/93/0532–0035$06.00/0

Particles and Polymers

Ceria Particles   The particles are nanometric crystallites of cerium oxide (ceria), produced by aqueous precipitation from a solution of cerium nitrate. The average mass of particles has been determined from the intensity of scattered light; the value is $M_W = 2 \ 10^5$ g/mol. The average diameter has been determined from QELS, i.e. fluctuations in the intensity of scattered light ; the value is $\sigma = 60$ Å. The internal structure yields broad lines in X ray diffraction patterns, and the positions of these lines match the lines of ceria ; the size of crystallites was calculated from the width of these lines ; it was found to be comparable with the particle size. Neutron diffraction patterns show that some of the hydration water also contributes to the diffraction lines; hence the hydrated surfaces of the particles must be crystalline, i.e. each water-particle interface is a hydrated crystal face. The specific surface of the particles can be calculated from the Porod limit of small angle X ray scattering ; this yields A = 400 m^2/g ; BET adsorption experiments yield a lower value. All these data are consistent with a model where each particle is a nanometric crystallite of ceria with the shape of a platelet. High resolution MET confirms this model (**Figure 1**). A droplet of dilute dispersion has been spread on a collodion membrane and then dried. So the picture doesn't reflect the state of dispersion but gives an idea of the morphology of the particle. The black lines are given by the diffraction from 111 planes which are oriented normal to the membrane.

In water the surface groups of ceria tend to ionize according to acid-base equilibria:

$$Ce\text{-}OH_2^+ + H_2O \quad \longleftrightarrow \quad Ce\text{-}OH + H_3O^+$$
$$Ce\text{-}OH + H_2O \quad \longleftrightarrow \quad Ce\text{-}O^- + H_3O^+$$

The point of zero charge point is around pH=7.6 (*3*) ; at lower pH both equilibria are shifted to the left hand side, and the surfaces are positively charged. We have worked in the pH range from 1 to 3.4. The density of surface charges depends on pH, and its magnitude is not known ; however this is not important for interparticle forces ; indeed most surface charges are compensated by NO_3^- counterions which condense on the surface. The net charge Z^{eff} only depends on the magnitude of the particle diameter σ (*4-5*)

$$Z^{eff} = 2\sigma/L_B \tag{1}$$

where L_B is the Bjerrum length which determines the strength of ionic interactions in water. For the ceria particles it leads to $Z^{eff} = 17$ electron charges per particle.

These surface charges generate electrostatic repulsions between the particles. We have determined these repulsions according to the second virial coefficient of the dispersion. This was done in the following way. A sample at a concentration C_p of 20 g/l was washed in an ultrafiltration cell by HNO_3 solutions at fixed pH. Then dilutions were made at the same pH ; no other ions were added, i.e the ionic strength was set by pH. Finally the intensity of scattered light was measured for all dilutions at a given pH, and the second virial coefficient v was determined as the slope of c/I vs c. This second virial coefficient measures the excluded volume around a particle ; hence the range σ_r of electrostatic repulsions is calculated as :

$$\frac{4}{3} \pi \sigma_r^3 = v \tag{2}$$

The range was found to increase with pH :

$$pH = 1,7 \qquad \sigma_r = 140 \text{ Å}$$
$$pH = 2,5 \qquad \sigma_r = 200 \text{ Å} \qquad\qquad (3)$$
$$pH = 3,4 \qquad \sigma_r = 276 \text{ Å}$$

PAM Macromolecules The macromolecules were made of poly (acrylamide), thereafter abbreviated PAM. They were synthesized according to a free radicalar polymerisation process. We have used two types of macromolecules. Their average molar mass are $M_w(1) = 1.5 \ 10^5$ g / mol and $M_w(2) = 6 \ 10^5$ g / mol, and their mass distribution index are $M_w/M_n(1) = 2$ and $M_w/M_n(2) = 4$.

In the conditions of our experiments the macromolecules are uncharged ; at lower pH the amide group would be protonated ; at much higher pH and high temperature it could be hydrolyzed. Thus we have a neutral polymer which carries large dipoles, and it is in a good solvent .

Interaction between Particles and Polymers

Adsorption mechanism PAM macromolecules adsorb on ceria, as they do on other transition metal oxides (6). In the pH range 1-7 the general mechanism is H bonding from electronegative atoms on the polymer to acid protons on the surface ; its strength is determined by the relative acidity / basicity of chemical groups on the polymer relative to the oxide. On the polymer, carbonyls are available for bonding through their spare electron doublets ; amide nitrogens also carry an electron doublet, but at low pH these are protonated and unavailable for bonding.

Adsorption isotherms The amount of polymer adsorbed on the particles was determined through the depletion method. PAM macromolecules were equilibrated for 24 hours with a ceria dispersion, then the particles and adsorbed polymer were removed through ultracentrifugation and the amount of polymer remaining in the supernatant was measured with a total organic carbon analyzer. This method requires 2 hours of centrifugation at 50000 rpm ($1,5 \ 10^5$ g) to achieve a good separation between particles and free polymer ; indeed under these conditions it was verified that all particles go to the bottom while all free macromolecules remain in the supernatant.

Figure 2 shows the adsorption isotherm the polymer on the dispersion at two different pH.

There are two important points. First, the plateau of adsorption is reached when the free polymer concentration is $3 \ 10^{-3}$ g/cm^3. This value is relatively high and is related to a low affinity of the polymer for the surface.

Secondly, the value Γ_{max} of the plateau varies from 0.8 g of polymer per g of ceria at pH 1.to 1.6 g/g at pH 2.5. The quantity of polymer at the plateau is of the same order as the quantity of particles and this shows that the dispersion must be considered as a two components system.

It is still possible to express these results in the usual form of polymer coverage per unit surface area. With a surface area of 400 m^2/g, the plateau is obtained at 2.5mg of PAM per m^2 of ceria. This is comparable to the value of the PEO-SiO2 system (7) where the plateau is at 1mg/m^2 if we take into account the difference between the monomer mass of each polymer, 44 for the PEO and 71 for the PAM. But because there is some uncertainty on the specific surface of the dispersion, the results are best expressed in mass of bound polymer per mass of particles.

10 nm

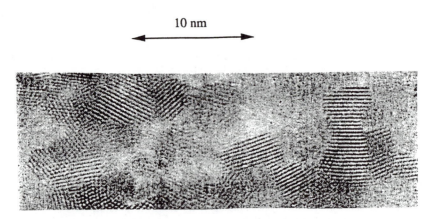

Figure 1. High resolution electron microscopy of ceria particles deposited through a drying method from a drop of dispersion onto a collodion membrane.

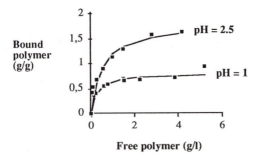

Figure 2. Adsorption isotherms for poly(acrylamide) macromolecules onto ceria particles. Horizontal axis : concentration of free polymer in the supernatant. Vertical axis: amount of bound polymer for 1g of ceria.

General Behavior of Attractive Two-Component Systems

The results from adsorption isotherms indicate that the dispersions may be considered as a two-component system. The relevant case is an A-B solvent system with A-A and B-B repulsions but A-B attractions. The general behavior of such systems features a phase separation region where concentrated A-B phase separates from a dilute one, and no separation when there is a sufficient excess of either A or B (*8*). This is easy to visualize in the present case : indeed saturation of the particles surfaces with polymer should lead to a steric repulsion between particles, and similarly saturation of the macromolecules with bound particles may also results in objects which are unable to bind to each other. Macroscopic flocculation is known to occur with surfaces which are covered with 1/3 to 1/2 of their saturation cover, hence this is also considered with the two-component description.

However this description is based on composition only, and it assumes that for one given composition the system will always reach a unique equilibrium state. This may not be the case, because polymer adsorption on a particle may be irreversible if sufficiently many monomers are bound. Thus the path followed to prepare samples may be important, and depending on this path many different non equilibrium states could be. This question of approach to equilibrium is discussed next.

Thermodynamic Equilibrium

How can we ascertain whether or not the dispersion is in thermodynamic equilibrium? Three criteria may be used.

(a) Time dependence of the "final" state : if the "final" state still evolves with time, then it is not the equilibrium state. However this criterion is worthless in the reverse direction : stability may be caused by high free energy barriers, and does not imply thermodynamic equilibrium.

(b) Comparaison of the final states obtained through different routes : If the same final state is obtained for samples of identical composition but prepared through different paths across the phase diagram, then this state has a good chance of being unique, and for all pratical purposes it can be treated as an equilibirum state. This was done in previous work on the same CeO_2+PAM dispersions (Spalla O, Cabane B. *Coll. Pol. Sci.*, in press).

(c) Reversibility in the processes which lead to the final state. If all forward reactions are balanced by reverse reactions which are efficient on the time scale of the experiment, then again the final state has the properties of an equilibrium state.

The last criterion has been used in the present work. The reaction of interest are bridging processes which lead to flocculation of the sample. Initially the sample was prepared at a composition where these bridging forces are opposed by electrostatic repulsions which prevent flocculation. The concentration of polymers was 0.5 g/l and that of particles 2.6 g/l, and the ionic strength was set by pH which was initially of 3.4. Then acid was added to the dispersion in order to bring the pH down to 1.7. The range of repulsion diminished and a macroscopic flocculation occured.

This was the "forward" reaction ; then reversibility could be investigated by investigating the "reverse" reaction i.e. lowering the ionic strength back to its initial value and examining whether the disperson would return to a stable sol. Accordingly, the flocculated dispersion was dialysed in order to raise the pH to its inital value of 3.4. In a few days (4 days) the sample was again stable.

From these two successive experiments it can be concluded that (**i**) flocculation can be induced by macromolecules (**ii**) this flocculation is a reversible process. The reversibility is due to the low energy of interaction between both components. This last point was obvious from the adsorption isotherm (**Figure 2**), where significant amounts of free polymer remain in equilibrium with unsaturated surfaces.

Reversibility of flocculation does not imply that the floc is in full equilibrium. Indeed the time scale for reverse reactions may be short in local reordering processes but long in large scales ones. Still it does ensure that the experimental boundary between one-phase and two-phase samples is an equilibrium boundary : out of equilibrium flocs will redissolve. Accordingly it is possible to determine a phase diagram.

Phase Diagram

Phase diagram determination Macroscopic separations occured at pH 1.7. Hence for this pH a phase diagram was constructed according to the composition of separating phases. Samples were made at fixed concentration of particles and increasing concentrations of polymer. Then, visual observations were used to classify the samples. Some samples remain one phase and flow under shear : they are classified as sols. Others samples are one phase but they doesn't flow under shear : they are classified as gels. The last types of samples are phase-separated and they define the two phases region.

In a second step, we foccussed on the two phase region. The samples separate in two phases : one is dense and sediments with time leaving the supernatant which constitutes the second phase. The two phases may contain ceria particles and polymer and it is necessary to know the exact composition of the two phases. In the supernatant the ceria particles are titrated through cerium atomic adsorption and the polymer through total organic carbon determination. The concentration of particles in the floc is determined through ceriumIV chemical oxido-reduction processes. The results are reported on **Figure 3**.

Main Features of the Phase Diagram The phase diagram presents very instructive features. First, the upper limit of the two phase region is a straight line giving a composition limit of 0,8g/g. This value is to be compared with the plateau of the isotherm of adsorption. This only reflects the stabilization due to steric repulsions between particles. The lower limit corresponds to the saturation of macromolecules by particles. This limit has been analyzed in a previous paper (Spalla O, Cabane B. *Coll. Pol. Sci.*, in press).

When entering the two phase region through the upper limit the samples separate into a supernatant which contains a large fraction of particles and a floc which is not very concentrated. This is not good for industrial application. When entering through the lower limit the floc is more concentrated but still leaves a great number of particles in the supernatant. This is not better. But there exists a composition for which the supernatant contains the least particles and for which the floc is the most concentrated. This means that the floc sediments easily in the sample and that the supernatant is clear. This is exactly what one usually calls an Optimum of Flocculation. This is a very good point for polymers because it shows that they are able to efficiently clean water from nanometric particles.

Finally all tie lines are of slope near one in a log-log diagram. This means that the composition of the floc is nearly the same as the composition of the supernatant.

Influence of the polymer mass on the phase diagram A phase diagram was constructed with another polymer of molecular weight ($M_w = 150\ 000$ g/mol). The general layout of the phase diagram is not affected by the change of macromolecules (**Figure 4**). There are differences only for dilute samples. The tip at low concentrations is less important in the case of low molecular weight macromolecule. This is due to the entropy of dispersion which rises with the number of objects in dispersion.

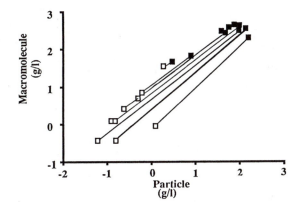

Figure 3. Phase diagram for dispersions of ceria particles (molar mass 200 000g/mol) bound to poly(acrylamide) macromolecules (molar mass 600 000g/mol). Open squares: composition of the supernatant, filled squares : composition of the corresponding flocs. Lines rely flocs and their supernatants ; they don't exactly correspond to the tie lines because of the log-log representation.

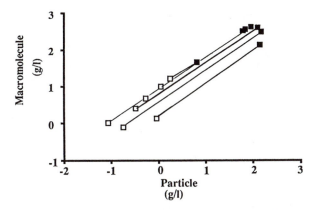

Figure 4. Phase diagram for dispersions of ceria particles (molar mass 200 000g/mol) bound to poly(acrylamide) macromolecules (molar mass 150 000g/mol). Open squares: composition of the supernatant, filled squares : composition of the corresponding flocs.

Criticism of the phase diagrams The phase diagrams appear consistent with expectations from the theory of mixtures, and reversibility experiments show that their boundaries are equilibrium boundaries. Still, the locations of these boundaries may be inaccurate for 2 reasons. First, the two-phases (sol and floc) may not separate properly and one of them may take an excessive amount of one component. We have checked this by making sols located at the boundary through direct mixing, and we have found that the location of the "sol" boundary is accurate. Second, the volume measured for the flocs may not reflect their concentrations. Indeed, the flocs are turbid, which indicates that they must be heterogeneous, i.e. made of lumps and voids.

Obviously the voids in the flocs are not significant for the thermodynamics of the two-component dispersions. Hence, proper phase diagrams must reflect the concentrations in flocs which contain no voids. For this purpose the flocs must be compressed.

Osmotic Compression

Instead of applying a mechanical pressure on the samples we used a method introduced by A. Parsegian (9-10) ten years ago. Samples were placed in dialysis bags and these bags were immersed in a large reservoir at high osmotic pressure. The osmotic pressure in the reservoir was set by the concentration of other macromolecules. The membranes of the bags were impermeable to the particles and to both types of macromolecules. On the other hand small ions and water molecules could diffuse through the membranes. As the osmotic pressure of macromolecules in the reservoir was large, water molecules were drawn from the bags and the samples were compressed. In our case the osmotic pressure was set at five atmospheres. After osmotic equilibirum was reached the concentrations inside the bags were measured again. The new concentrations are reported **Figure 5**.

They are two interesting features in this phase diagram : **(i)** Along the upper boundary where the concentrations of separating phases are not very different, the shift obtained through osmotic compression of the floc is rather small. This shows that such samples are fairly well ordered and contain few voids. **(ii)** For mixtures made right in the middle of the two-phases region, which separate into dilute sols and concentrated flocs, the shift caused by compression of the flocs is quite large. This shows that flocs made near the O.F.C are poorly ordered and contain many voids ; the overall concentration of such uncompressed flocs has no thermodynamic significance.

It is also interesting to remark that the magnitude of the shift follows the composition : flocs made of nearly saturated particles contain few voids, whereas flocs made of highly unsaturated particles contain many more voids. This is presumably related to the larger strength and shorter range of bridging attractions between unsaturated surfaces which do not allow reordering process to occur spontaneously.

Still one may wonder how much this shift depends on the applied pressure. Indeed the new concentrations were measured under an osmotic stress of 5 atmospheres ; would they have been substantially different if the applied pressure was more or less? This requires a measurement of the complete osmotic compression curve for flocs.

Equation of State for Flocs and Gels

The complete osmotic compression curve was determined through a set of "Parsegian" experiments. At equilibrium the osmotic pressures inside and outside the bag are equal. Thus when the concentration of macromolecules in the reservoir is raised the osmotic pressure in the bag increases as well. The response of the dispersion inside the bag is a further deswelling. In this way the concentration inside the bag was measured at different osmotic pressures for two compositions of the dispersion. The first composition is the O.F.C and the other one corresponds to the saturation of the

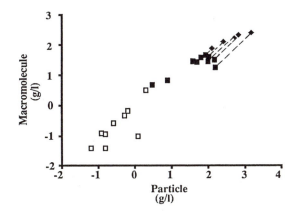

Figure 5. Phase diagram for dispersions compressed through osmotic stress.
Open squares : compositions of the supernatants
Filled squares : compositions of the flocs
Filled diamonds : compositions of the compressed flocs.

surfaces. The curves of osmotic pressure versus the particle volumic fraction are reported in **Figure 6**.

At the O.F.C two different regimes are found. First, for particles concentrations below 20%, the compression is quite easy. Then at concentrations higher than 20% the pressure undergoes a very sharp rise. The fact that at pressures higher than a few atms, the concentration does not change with pressure indicates that the previous measurements under an osmotic stress of five atmospheres were correct.

How can we interpret the two regimes of pressure versus concentration? In the first regime the compression is easy because it corresponds tò the elimination of voids. This is interesting because it shows that it is possible to eliminate the voids. So it means that particles are allowed to roll around each other to rearrange. This is because polymers lubricate the contacts between particles. Nevertheless it takes an osmotic pressure of one atmosphere at least to eliminate these voids and this is very instructive. Indeed, when the floc is in equilibrium with its supernatant, the osmotic pressure inside the floc is the same as in the supernatant. The osmotic pressure of the supernatant can be easily determined because it is dilute. It is around 10Pa, far less than the pressure required to eliminate voids. So the elimination of the voids cannot be a spontaneous process and this is why flocs are in general turbid and characterized as heterogeneous.

Then, in the second regime, when the voids have been eliminated, the pressure rises sharply. This is because the sample is made of a dense packing of particles covered with unsaturated layers. The conclusion is that the intrinsic structure of flocs is a homogeneous gel, and that voids are an extrinsic feature resulting from the flocculation process. This justifies the construction of the phase diagrams with a two-phase region surrounded by a unique one-phase region.

At the other composition, with saturated surfaces, the pressure rises much earlier with volume fraction. Denses packing may be defined as the volume fraction for which the pressure curves becomes nearly vertical. This is obtained at a volume fraction of 0.05 for saturated layers, and 0.2 for unsaturated layers. This shows that the effective hard core which has to be considered in the dense packing is the sum of the hard core of the particle plus the polymer layer thickness. This is in good agreement with the idea that a polymer layer is essentially incompressible. A saturated layer is thicker than an unsaturated one and this leads to a lower concentration in the case of floc compared to nearly saturated particles.

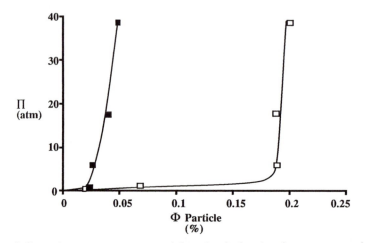

Figure 6. Osmotic pressure versus particle volumic fraction for two compositions of the dispersion. Open squares : O.F.C., filled squares : saturated surfaces. The lines don't correspond to fits but are there to guide eyes

Conclusion

Dispersion of nanometric particles and adsorbing polymers behave like reversible 2-component systems (A-B-solvent). Under conditions where attractions overcome repulsions there is a phase separation into a gel and a supernatant which have nearly the same polymer/particles ratio. Concentrated gels can be made easily through phase separation in the 2-phases region ; however they are full of voids. These voids have been eliminated through compression and the equations of state of gels with different polymer/particles ratio have been determined.

Literature Cited

1. Verwey, E.J.W.; Overbeek, J.T.G.; Theory of the stability of lyophobic colloids; Elsevier: Amsterdam **1948**.

2. Napper,D.H.; Polymeric Stabilization of colloidal dispersions; **1983**.
3. Ray,K.C.; Sengupta,P.K.; Roy,S.K; *Indian Journal of Chemistry* **1978**, *17A*, 348.
4. Belloni,L.; Drifford,M.; Turq,P., *Chem. Phys.* **1984**, *83*, 147.
5. Belloni,L.; In *Neutron, X ray and light scattering*; Lindner,P., Zemb,Th., Ed.; North Holland, **1991**.
6. Lee, L.T.; Somasundaran,P.; *Langmuir* **1989**, *5*, 854.
7. Lafuma, F.; Wong, K., Cabane, B.; *J. Colloid Interface Sci.* **1990**, *143*, 9.
8. Thalberg, K.; Lindman, B.; Karlström, G.; *J. Phys. Chem.* **1990**, *94*, 10.
9. Parsegian, V.A.; Fuller, N.; Rand, R.P.; *Proc. Natl. Acad. Sci. U.S.A.* **1979**, *76*, 2750.
10. Parsegian, V.A.; Rand, R.P.; Fuller, N.; *J. Phys. Chem.* **1991**, *95*, 4777.

RECEIVED December 22, 1992

Chapter 5

Pairwise Interactions in the Critical Micelle Concentrations of Diblock Copolymers

Yongmei Wang[1], Wayne L. Mattice[1], and Donald H. Napper[2]

[1]Institute of Polymer Science, University of Akron, Akron, OH 44325–3909
[2]Department of Physical and Theoretical Chemistry, School of Chemistry, University of Sydney, New South Wales 2006, Australia

Simulations on a cubic lattice have been used to study the self-assembly into micelles by diblock copolymers with N_A segments of block A and N_B segments of block B. Previous work has shown these simulations can be used to measure the critical micelle concentration (cmc). In the present work, the pairwise interactions cause the medium to behave as a poor solvent for block A and a good solvent for block B. The purpose is to determine the relative importance of the three pairwise interactions between different species, denoted by E'_{AS}, E'_{AB} and E'_{BS}, for the cmc. Here E'_{XY} denotes the dimensionless pairwise interaction, U_{XY}/kT, where U_{XY} is the pairwise interaction energy of segments of species X and Y. The three species involved here are segments of A, segments of B, or solvent (S). Hence we examine all E'_{XY} where X $\neq$ Y. The results are compared with previous work that focused attention on the dominant pairwise interaction parameter, E'_{AS}. The effects of E'_{AB} and E'_{BS} are less important for asymmetric $(N_B/N_A > 1)$ than symmetric $(N_B/N_A = 1)$ block copolymers, and are less important than the effect of E'_{AS} for all cases examined.

Diblock copolymers can self-assemble into micelles when immersed in a selective solvent of small molecules, or in a matrix comprised of a homopolymer. The micelles are easily detected by techniques such as light scattering (*1*) and transmission electron microscopy (*2*), but the critical micelle concentration (cmc) may be so small that other techniques are required for its determination. Techniques based on the ability to detect fluorescence from very dilute systems have been used for this purpose (*3*).

The theoretical formulations of the cmc of symmetric diblock copolymers identify the important parameter as χN_A, where χ is the interaction parameter

0097–6156/93/0532–0045$06.00/0

for a segment of the insoluble block and the solvent, and N_A denotes the number of segments in the insoluble block (4). Hence χN_A is the interaction per insoluble block. The prediction of the analytical theory for symmetric diblock copolymers under conditions of strong segregation,

$$\ln \text{cmc} \sim -\chi N_A \tag{1}$$

has been reproduced in a recent simulation of the self-assembly of diblock copolymers in dilute solution (5). The requirement that the diblock copolymers be symmetric could be relaxed, provided the cmc was expressed in terms of the volume fraction of the insoluble segments. That work focused on the influence of the interaction of the segments of the insoluble block with the other components of the system. The same simulation has also been used successfully for the study of the self-assembly of symmetric triblock copolymers (6).

Here we turn to a consideration of the influence of two other pairwise interactions in the diblock copolymer, namely the interactions between the segments in the two blocks (which is assumed to be repulsive), and the interaction of the soluble block with the solvent (which is assumed to be attractive). The importance of the variation in these two terms, and the connection with the symmetry of the diblock copolymer, are deduced from the simulations.

Method

The simulation is performed on a cubic lattice of dimensions L^3, $L = 44$, with periodic boundary conditions in all directions (5). Reptation (7) and the extended Verdier-Stockmayer moves (8) are used to convert one replica into another, with the Metropolis rules (9) employed for acceptance of new replicas. Each of the 200 diblock copolymers contains N_A beads of A and N_B beads of B. Vacant lattice sites are considered to be occupied by solvent, S.

The system is subject to two types of energies, one type of which is constant. Each bead of A or B has a constant hard core, such that no site on the lattice may be occupied by more than one bead. There is also a variable pairwise interaction applied when X and Y are nonbonded nearest neighbors. These interactions are assigned reduced energies

$$E'_{XY} = \frac{U_{XY}}{kT} \tag{2}$$

where U_{XY} is the internal energy for the pairwise interaction of components X and Y. Each simulation begins with a dispersed state, and is run to equilibrium with the desired set of E'_{XY}. A diblock copolymer is defined as a free chain if none of its N_A beads of A has a nearest neighbor that is an A bead from another diblock copolymer. Those chains that are not free are defined as participating in aggregates. Two-dimensional projections of representative micelles formed above the cmc when $N_A = N_B = 10$, $E'_{AS} = 0.45$, and $E'_{AB} = E'_{BS} = 0$ are depicted in Figure 1.

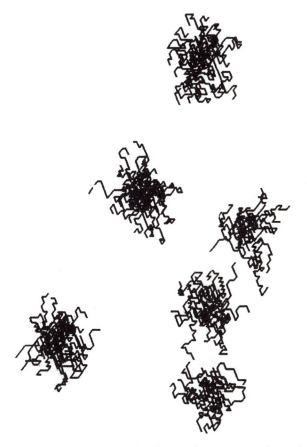

Figure 1. Two-dimensional projections of representative micelles when $N_A = N_B = 10$, $E'_{AS} = 0.45$, and all other pairwise interactions are zero. The black-and-white representation does not permit distinction between the two types of beads.

The cmc is determined in the simulations by counting the number of free chains, N_{free}, when the total concentration is high enough to promote extensive formation of large aggregates at equilibrium (5). The cmc will be expressed as the volume fraction of the diblock copolymer that is present as free chains after equilibrium is established

$$V_{AB}^{cmc} = \frac{N_{free}(N_A + N_B)}{L^3} \tag{3}$$

or as the volume fraction of the insoluble A block in these same free chains

$$V_A^{cmc} = \frac{N_{free}N_A}{L^3} \tag{4}$$

Results and Discussion

Summary of Previous Results for the Influence of $E'_{AS} = E'_{AB} > 0$. The role of the repulsive interaction of the insoluble block with its surroundings, be they solvent or the soluble block, was studied recently (5). The cmc's obtained for 18 systems, with V_A^{cmc} covering a twenty-fold range, $0.00058 < V_A^{cmc} < 0.013$, yield a straight line with slope -1 when plotted as $\ln V_A^{cmc}$ vs. $E'_{AS}N_A$.

$$\ln V_A^{cmc} \sim -E'_{AS}N_A \tag{5}$$

This result, obtained with diblock copolymers for which $1 \le N_B/N_A \le 7$, is consistent with Leibler's prediction for the case of strong segregation and $N_B/N_A = 1$. It shows that the cmc is strongly dependent on the repulsive interaction of the insoluble block, expressed as the product of the energy per segment, and the number of segments.

One might surmise that the more important of the two parameters varied in that study was E'_{AS}, because there are likely to be more contacts between A and solvent than between A and B in dilute solution. The validity of that surmise is examined in the next section. Then we will turn to the influence of E'_{BS}.

The Role of E'_{AB}. The previous simulations employed $E'_{AS} = E'_{AB}$. Here we break this equality in order to evaluate the importance of the repulsion between the two blocks, represented by E'_{AB}. The cmc deduced from simulations with $E'_{AB} > 0$ and $E'_{AS} = 0.45$ are presented in Table I. The value of E'_{AS} must be positive if micelle formation is to occur in dilute solution. After equilibrium is established there is a fluctuation in N_{free} due to the exchange of diblock copolymers between the aggregates and the pool of free chains (5). This fluctuation in N_{free} corresponds to about ± 0.0007 for the V_A^{cmc} evaluated by Equation (4).

The imposition of a repulsive pairwise interaction when segments of A and B occupy adjacent sites plays an interesting role. This role is most easily understood by focusing on the cmc as measured by the volume fraction of the insoluble block, V_A^{cmc}. From the first and last lines in Table I, we note that a tripling in the size

Table I. Simulations for $A_{10}B_{10}$ and $A_{10}B_{30}$ with E'_{AB} Variable[a]

| E'_{AB} | $N_B/N_A = 1$ | | $N_B/N_A = 3$ | |
	V_A^{cmc}	V_{AB}^{cmc}	V_A^{cmc}	V_{AB}^{cmc}
0.0	0.0018	0.0035	0.0040	0.016
0.1	0.0027	0.0054	0.0041	0.016
0.3	0.0036	0.0073	0.0046	0.018
0.45	0.0042	0.0085	0.0045	0.018
0.7	0.0045	0.0089	0.0043	0.017

[a] Using $E'_{AS} = 0.45$, $E'_{BS} = 0$.

of the soluble block will double the cmc if $E'_{AB} = 0$, but produces no effect if $E'_{AB} = 0.7$. This result can be rationalized by remembering the simulation always prohibits double occupancy of any lattice site by segments of A or B. The hard-core repulsion of A and B produces the difference in cmc when $E'_{AB} = 0$. Even when the nearest neighbor interaction of A and B is turned off, the excluded volume interaction between the soluble blocks makes it more difficult to create micelles with long soluble tails than with short soluble tails. This effect is entropic in origin, because it arises from the reduction in the number of configurations accessible to the soluble block in diblock i due to the excluded volume interaction with the other diblock copolymers in the micelle. The effect would be reduced if the soluble blocks were less flexible, because they would then have less entropy to lose. The effect could also be reduced if the soluble blocks tended to avoid the domain of the insoluble blocks even in the free chains, because there would be less entropy to lose when the free chains were incorporated into micelles. Turning on the repulsive interaction between A and B causes the two blocks to avoid one another in the free chain, and therefore reduces the impact of the lengthening of the soluble block on the cmc.

From the second and fourth columns in Table I, we see that increasing E'_{AB} from 0 to 0.7 more than doubles the cmc if $N_B/N_A = 1$, but produces a negligible effect if $N_B/N_A = 3$. The explanation for this observation is closely tied to the phenomenon described in the preceding paragraph. As the soluble block increases in size, the hard-core excluded volume interaction in the free chain will cause the beads in the soluble block to more strongly avoid the beads in the insoluble block, even when the nearest neighbor interaction is nil. Therefore turning on the repulsive nearest neighbor interaction, by making $E'_{AB} > 0$, has a diminishing effect on the cmc as the size of the soluble block increases.

Finally, we can deduce that the effect of E'_{AB} is much smaller than the effect of E'_{AS}. The variation of E'_{AB} in Table I never changes V_A^{cmc} by as much as a factor of three, but variation of $E'_{AS} = E'_{AB}$ over a somewhat smaller range produced changes in V_A^{cmc} of more than a factor of twenty (5). This result is in conformity with the surmise detailed above and arises from the higher probability of AS contacts.

Table II. Simulations for $A_{10}B_{10}$ and $A_{10}B_{30}$ with E'_{BS} Variable[a]

| E'_{BS} | $N_B/N_A = 1$ | | $N_B/N_A = 3$ | |
	V_A^{cmc}	V_{AB}^{cmc}	V_A^{cmc}	V_{AB}^{cmc}
0.0	0.0041	0.0082	0.0045	0.018
-0.1	0.0047	0.0094	0.0045	0.018
-0.2	0.0048	0.0096	0.0045	0.018
-0.3	0.0053	0.0106	0.0041	0.016
-0.4	0.0053	0.0106	0.0056	0.023
-0.5	0.0054	0.0108	0.0040	0.016

[a] Using $E'_{AS} = E'_{AB} = 0.45$.

The Role of E'_{BS}. Next we describe the role of E'_{BS} when $-0.5 \leq E'_{BS} \leq 0$, which corresponds to the medium being a good solvent for block B. The values of E'_{AS} and E'_{AB} are both 0.45, which corresponds to the medium being a poor solvent for block A (which is necessary for micelle formation), and a repulsion between blocks A and B. Table II presents the results for a series of simulations performed with symmetric diblock copolymers that have $N_A = N_B = 10$. The fluctuations in the system after equilibrium is established correspond to about ± 0.0007 for V_A^{cmc} and ± 0.0014 for V_{AB}^{cmc}. The cmc increases slightly as the solvent provides a better medium for B, but the effect is small and approaches a limiting values as E'_{BS} becomes negative. The changes in the cmc barely exceed the ranges that accompany the fluctuations of ± 0.0007 for V_A^{cmc}.

The last two columns in Table II present the results for an asymmetric diblock copolymer in which the soluble block is three times longer than the insoluble block. The influence of E'_{BS} on the cmc is even smaller for the asymmetric diblock copolymer than for the symmetric diblock copolymer. When compared with the previous work (5), where E'_{AS} was varied, we find that the influence of E'_{BS} on the cmc is trivial compared with the influence of E'_{AS}.

When the two diblock copolymers are compared in the same solvent, the cmc is larger for the asymmetric diblock than for the symmetric diblock if it is expressed as V_{AB}^{cmc}, but there is little change if the cmc is expressed as V_A^{cmc}. The earlier study (5) also concluded that V_A^{cmc} was the more informative method of expressing the cmc of the diblock copolymers with variable N_A and N_B. That study also found V_A^{cmc} was proportional to $E'_{AS}N_A$. Here $E'_{AS}N_A$ gives the differences in energy for an A block completely immersed in the core of a micelle and completely surrounded by solvent. Clearly there is no useful purpose in attempting to relate V_A^{cmc} to $E'_{BS}N_B$. The results also suggest that improvement of the solvation of the soluble block does not modify the form of the dominating influence of the insoluble block.

Interplay of E'_{AB} and E'_{AS}. Usually we expect $E'_{AB} > 0$, and that condition was used in Table II. If this nearest neighbor repulsion between segments in unlike blocks is turned off, E'_{BS} can play a more important role in the determination of the cmc, as shown in Table III. This Table repeats selected information from

Table III. Simulations for $A_{10}B_{10}$ with E'_{BS} Variable and Two Different Values of $E'_{AB}{}^a$

| | $E'_{AB} = 0.45^b$ | | $E'_{AB} = 0$ | |
E'_{BS}	V_A^{cmc}	V_{AB}^{cmc}	V_A^{cmc}	V_{AB}^{cmc}
0.0	0.0041	0.0082	0.0018	0.0035
-0.1	0.0047	0.0094	0.0029	0.0058
-0.4	0.0053	0.0106	0.0047	0.0094
-0.5	0.0054	0.0108	0.0049	0.0099

^{a}Using $E'_{AS} = 0.45$.
^{b}The data for $E'_{AB} = 0.45$ are from Table II.

Table II for $A_{10}B_{10}$, computed with $E'_{AB} = 0.45$, and compares it with results obtained when $E'_{AB} = 0$. For the range of E'_{BS} covered in both sets of simulations, the cmc varies by a factor of 1.3 if $E'_{AB} = 0.45$, but the variation in cmc increases to a factor of 2.7 if $E'_{AB} = 0$. The influence of E'_{BS} increases as the repulsion between the two types of blocks decreases.

Summary of the Importance of the Three E'_{XY} for the cmc. In determining the values of V_A^{cmc} under the conditions where the two blocks have repulsive nearest neighbor interactions, the pairwise energies with $X \neq Y$ can be ranked in importance as

$$E'_{AS} > E'_{AB} > E'_{BS} \qquad (6)$$

The first inequality derives from the relatively low probability of AB contacts in dilute solution whereas the second results from the relatively small change in the number of BS contacts upon self-assembly into micelles. The inequality becomes

$$E'_{AS} > E'_{AB}, E'_{BS} \qquad (7)$$

when the repulsive nearest neighbor interaction between the two blocks is turned off.

This chapter does not address other properties of interest, such as the distribution of the sizes of the micelles at equilibrium. Previous simulations have shown N_{free} equilibrates much faster than the weight average number of chains in the aggregates, M_w/M_0 (5). The length of the simulation required for accurate evaluation of N_{free} depends on the rate at which the free chains equilibrate with aggregates of all sizes. In contrast, an accurate evaluation of M_w/M_0 cannot be achieved until all of the aggregates, including the largest, have equilibrated with one another. That process may require an order of magnitude more computer time than the equilibration of N_{free}. Preliminary estimates for M_w/M_0 suggest it is affected by the interaction energies, as expected. The preliminary results also suggest E'_{BS}, in the range covered here, may have a more important effect on the kinetics than on the cmc.

Acknowledgment

This research was supported by the National Science Foundation, through grants DMB 87–22238 and INT 90–14836, and by the Australian Department of Industry, Technology and Commerce.

Literature Cited

1. Tanaka, T.; Kotaka, T.; Inagaki, H. *Polym. J.* **1972**, *3*, 338.
2. Kinning, D. J.; Winey, K. I.; Thomas, E. L. *Macromolecules* **1988**, *21*, 3502.
3. Zhao, Y. L.; Winnik, M. A.; Riess, G.; Croucher, M. D. *Langmuir* **1990**, *6*, 514.
4. Leibler, L.; Orland, H.; Wheeler, J. C. *J. Chem. Phys.* **1983**, *79*, 3550.
5. Wang, Y.; Mattice, W. L.; Napper, D. H. *Langmuir*, in press.
6. Wang, Y.; Mattice, W. L.; Napper, D. H. *Macromolecules* **1992**, *25*, 4074.
7. Rodrigues, K.; Mattice, W. L. *J. Chem. Phys.* **1991**, *94*, 761.
8. Verdier, P. H.; Stockmayer, W. H. *J. Chem. Phys.* **1962**, *36*, 227.
9. Metropolis, N.; Rosenbluth, A. N.; Rosenbluth, M. N.; Teller, A. H.; Teller, E. *J. Chem. Phys.* **1953**, *21*, 2087.

RECEIVED December 22, 1992

Chapter 6

Interactions of Globular Colloids and Flexible Polymers

Understanding Protein Partitioning in Two-Phase Aqueous Polymer Systems

Nicholas L. Abbott[1], Daniel Blankschtein, and T. Alan Hatton

Department of Chemical Engineering, Massachusetts Institute of Technology, Cambridge, MA 02139

We review our recent work aimed at elucidating the molecular-level mechanisms responsible for the partitioning of globular proteins in two-phase aqueous poly(ethylene oxide) (PEO)-dextran systems using the theoretical tools of polymer-scaling concepts, statistical-thermodynamics, and liquid-state theory, as well as the complementary experimental techniques of equilibrium protein partitioning and small-angle neutron scattering (SANS).

On the basis of scaling predictions and SANS measurements, we propose that certain experimentally observed protein partitioning behaviors arise from a crossover in the underlying structure of the PEO-rich solution phase from individually dispersed PEO coils to an extensively entangled PEO mesh. Accordingly, we have introduced a variety of molecular-level pictures to describe the interactions of globular proteins and flexible polymers in solution. The various physical pictures differ in the ways they incorporate (i) the polymer solution regime, (ii) the relative size of the protein and the polymer coil/mesh, and (iii) the nature of the energetic interaction between the flexible polymer chains and the globular protein molecules.

We have explored each proposed physical picture through a scaling-thermodynamic formulation and a combined equation-of-state/Monte-Carlo approach, and predicted the associated protein partitioning behaviors. A comparison of the theoretical predictions with experimental protein partitioning measurements (using two-phase aqueous PEO-dextran systems or a diffusion cell) suggests that although the physical exclusion of the proteins by the polymers contributes to the observed protein partitioning behavior, other interactions also play a significant role. Specifically, we have predicted the existence of a weak attractive interaction between the polymers and the proteins which increases with protein size, R_p, where $17\text{Å} < R_p < 51\text{Å}$, from order $0.01kT$ to $0.1kT$ (per EO segment at the protein surface). The net interactions (physical exclusion and attraction) between the proteins and the PEO coils, however, remain

---

[1]Current address: Department of Chemistry, Harvard University, Cambridge, MA 02138

0097–6156/93/0532–0053$06.00/0

strongly repulsive. This view is corroborated using SANS to measure
the average spatial correlations (solution structure) between bovine
serum albumin (BSA, R_p=35Å) and singly dispersed PEO coils in
D_2O.

The interactions of globular colloidal particles and flexible chain macromolecules
control diverse phenomena such as the formation of complexes between polymers and
micelles (1-8), as well as the polymeric stabilization and flocculation of gold sols (9),
ceramic particles (10), and other colloidal dispersions (11). The macroscopic
manifestations of these interactions play a central role in the technologies associated
with photography, food processing, emulsion/microemulsion polymerization, and
enhanced oil recovery. Another example, which is explored in this paper, is aqueous
solutions of globular proteins and flexible (synthetic and biological) polymer molecules
under conditions where the polymer solution has undergone phase separation (12,13).
In the resulting *two-phase aqueous polymer system*, proteins usually distribute
unevenly between the two coexisting polymer solution phases. Investigations of the
partitioning of proteins and other biomolecules in two-phase aqueous polymer systems
have been stimulated by the potential of these polymer solutions to provide immiscible,
yet protein compatible, liquid phases for the purification of proteins by liquid-liquid
extraction (12-15). Furthermore, the presence of polymers in a protein purification
process may provide other advantages related to polymer-enhanced protein refolding
(16), and the stabilization of proteins in product formulations. However, at the
molecular-level, the understanding of the interactions of globular proteins and flexible
synthetic polymers which are responsible for the potentially useful properties of these
solutions needs further development(17-19).
　　In this review, we address three broad questions:
　　(i) What factors constitute a physical basis to develop an understanding of
protein partitioning in two-phase aqueous polymer systems? Specifically, how
important are geometry and energetics in determining protein-polymer interactions?
　　(ii) Can one use scaling concepts from polymer physics (20) to rationalize some
of the macroscopic manifestations, such as protein partitioning behaviors, associated
with different protein-polymer interaction mechanisms?
　　(iii) Can one reach a unified theoretical description of both the molecular-level
structure and the thermodynamic properties of protein-polymer solutions, and verify
both aspects using experimental methods?
　　To understand the essential physics and develop a physical basis to describe
protein partitioning in two-phase aqueous polymer systems, polymer scaling concepts
(20) proved to be a powerful theoretical tool (21). For a variety of proposed physical
pictures, we have developed scaling-thermodynamic descriptions of protein
partitioning, and predicted the associated protein partitioning behaviors (21). When
used in conjunction with equilibrium partitioning experiments, these theoretical
predictions provided a useful tool to discern between a number of partitioning
mechanisms. Subsequent and more detailed treatments of possible partitioning
mechanisms exploited an equation-of-state/Monte-Carlo approach (22) to predict both
experimental trends in the equilibrium protein partitioning behavior, as well as small-
angle neutron scattering (SANS) from solutions of proteins and polymers (23).
Specifically, from a common physical description of protein-polymer interactions, we
have been able to predict both the thermodynamic and structural properties of aqueous
solutions containing proteins and polymers, as well as confirm these predictions using
experimental measurements. Below we review our recent theoretical and experimental
studies. For a detailed description of the results, the reader is referred to(21-25).

Experimental Observations on Protein Partitioning in Two-Phase Aqueous Polymer Systems

The distribution of proteins between the two coexisting phases of a two-phase aqueous polymer system is characterized by a partition coefficient, $K_p = c_{p,t}/c_{p,b}$, where $c_{p,t}$ and $c_{p,b}$ are the protein concentrations in the top (t) and bottom (b) polymer solution phases, respectively. Experimental investigations have revealed that a large number of factors influence protein partitioning, including the types of polymers *(12)*, their molecular weight and concentration *(26,27)*, protein size, conformation and composition *(26,28,31)*, salt type and concentration, and solution pH *(29,30)*. In this review, we focus on understanding the influence of polymer properties on the partitioning of proteins, in addition to certain properties of the proteins, such as their size and the nature of their energetic interactions with the polymers.

The important role of polymer properties, such as polymer molecular weight, on the partitioning of proteins can be illustrated in the two-phase aqueous poly(ethylene oxide) PEO-dextran system *(12)*. Both PEO and dextran are separately miscible with water in all proportions and, at low polymer concentrations, with each other. As the polymer concentrations increase, however, phase separation occurs, with the formation of a two-phase system with a top phase that is rich in PEO and a bottom phase that is rich in dextran. For example, the protein partition coefficients reported in Figures 1 and 2 were measured in two-phase systems with top phases containing 11% w/w PEO and 89% w/w water, and bottom phases containing 19% w/w dextran and 81% w/w water. Because the overall composition of this two-phase system was far beyond the critical point composition, the compositions of each of the coexisting phases were insensitive to changes in PEO molecular weight *(26)*. Typically, the concentration of the minor polymer species in each phase was less than 1% w/w.

In Figure 1, experimentally observed partition coefficients for a variety of globular proteins are presented as a function of PEO molecular weight *(26,27)*. The observed response of the protein partition coefficients to changes in the molecular weight of PEO reveals that:

(1) An increase in the molecular weight of PEO results in the partitioning of the proteins away from the top PEO-rich phase, and hence in a decrease in $\ln K_p$.

(2) The response of K_p to changes in PEO molecular weight is greatest at PEO molecular weights below approximately 10,000 Da, as reflected by the rapid decrease of $\ln K_p$ over this range of PEO molecular weights.

(3) The response of K_p to changes in PEO molecular weight is protein specific, and correlates with the protein size (see also Figure 2).

Experimental observations such as those presented in Figures 1 and 2 motivated the philosophy behind our theoretical formulation *(21)*. In particular, the correlation observed between the overall sizes of the proteins and their partitioning behavior suggested that a coarse-grained view of these systems, rather than an atom-by-atom account, may be sufficient to describe the nature of the interactions responsible for the observed partitioning behaviors of these proteins. This realization prompted our decision to pursue *(21,24)* a scaling account *(20,32-34)* of protein partitioning in two-phase aqueous polymer systems.

It is also important to emphasize that the rationalization of the protein partitioning trends reported in Figures 1 and 2 was simplified considerably by the judicious choice of experimental conditions *(21,25)*. Specifically, under the chosen experimental conditions of (i) negligible PEO concentration in the dextran-rich phase, (ii) negligible dextran concentration in the PEO-rich phase, and (iii) the essentially constant weight *fractions* of PEO and dextran in the coexisting polymer solution phases over the range of PEO molecular weights investigated, it was possible to relate the origin of the observed protein partitioning trends to the influence of the PEO molecular weight on the PEO-rich phase *(21,25)*.

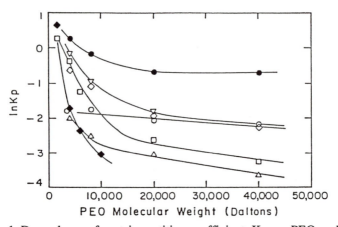

Figure 1. Dependence of protein partition coefficient, K_p, on PEO molecular weight in a PEO-dextran-water two-phase system. In order of increasing size the proteins are: (open circles) cytochrome-c, (filled circles) ovalbumin, (open diamonds) bovine serum albumin, (triangles up) lactate dehydrogenase, (triangles down) catalase, (open squares) pullulanase, (filled diamonds) phosphorylase. Data compiled from (26,27). The solid lines are drawn to guide the eye.

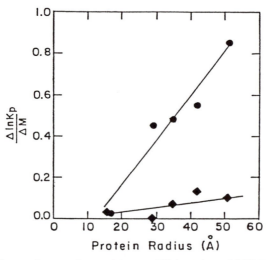

Figure 2. Change in protein partition coefficient (per 4,000 Daltons PEO), $\Delta\ln K_p/\Delta M$, with PEO molecular weight for five proteins; (filled circles) 4,000 Daltons to 8,000 Daltons, (filled diamonds) 20,000 Daltons to 40,000 Daltons. The data are taken from Figure 1. In order of increasing size the proteins are: cytochrome-c, ovalbumin, bovine serum albumin, lactate dehydrogenase and catalase. The solid lines are drawn to guide the eye.

Novel Physical Pictures of Protein Partitioning in Two-Phase Aqueous Polymer Systems

In order to describe the interactions between the proteins and PEO within the PEO-rich phase, the characteristic length scale of the PEO-rich solution phase was identified using two methods. First, the PEO concentration characteristic of the region where the extensive overlap of PEO coils begins to occur, c^*, was estimated *(21)* using the simple relation $c^* = 3M_2/4\pi R_g^3$, where M_2 and R_g are the molecular weight and radius of gyration of the PEO polymer molecules, respectively. A comparison of the calculated c^* versus M_2 curve to the measured PEO concentrations in the top PEO-rich phase revealed intersection of the two curves at PEO molecular weights of around 10,000 Da (see Figure 3),. This observation suggested that the protein partitioning behavior observed in Figures 1 and 2 reflects a *crossover in the structure of the PEO-rich solution phase*. This view of a transition in the nature of the PEO-rich phase was corroborated *(23)* by measurements, using SANS ($0.03\text{Å}^{-1} < q < 0.3\text{Å}^{-1}$, where q is the magnitude of the scattering vector), of the correlation length of the polymer solution (in D_2O) over a wide range of polymer concentrations (0.8% w/w to 25% w/w) and molecular weights (1,500 Da to 860,000 Da), see Figure 4. By interpreting these measurements in terms of a correlation length, ξ (which corresponds to the sizes of individual PEO coils in dilute polymer solutions, or to the mesh size ["blob size"] of a solution of extensively entangled PEO coils), a transition was observed *(23)* from the dilute to the entangled polymer solution regimes with increasing PEO molecular weight (at constant PEO weight fraction), and with increasing PEO weight fraction (at constant PEO molecular weight). Specifically, in solutions containing approximately 10% w/w PEO, the transition was observed in the vicinity of PEO 10,000 Da (see the vertical arrow in Figure 4). This observation supported our hypothesis *(21)* that changes in PEO molecular weight influence protein partitioning in the two-phase aqueous PEO-dextran system through an accompanying transition in the structure of the PEO-rich solution phase.

From the above considerations it was concluded that in systems containing low molecular weight PEO (M<10,000 Da), *individual polymer coils*, which may be larger or smaller than the proteins, interact with the proteins *(21)*. In contrast, in systems containing high molecular weight PEO (M>>10,000 Da), the proteins interact with an *entangled polymer mesh* rather than with identifiable polymer coils, and the protein partition coefficient becomes independent of PEO molecular weight *(21,24)*. The influence of the polymer solution regime on the nature of protein-polymer interactions is apparent in Figures 5 and 6, where a variety of different physical pictures are proposed to describe such interactions. Specifically, the nature of the protein-polymer interactions depends on (i) the polymer solution regime, (ii) the relative sizes of the protein and the polymer coil/mesh, and (iii) the nature of the energetic interactions between the flexible and diffuse polymer chains and the relatively rigid globular protein molecules *(21,24)*. Note that the consideration of energetic interactions was motivated, in part, by certain similarities between globular proteins and spheroidal ionic surfactant micelles, and the well-established role that energetic interactions can play in determining the properties of solutions containing micelles and polymers *(4-6)*.

Proteins in Solutions of Identifiable Polymer Coils.

For each of the physical pictures presented in Figure 5, we have developed mathematically-simple geometric and scaling arguments to probe the qualitative form of the free-energy change arising from protein-polymer and polymer-polymer interactions. Through a statistical-thermodynamic framework, used to relate this free-energy change

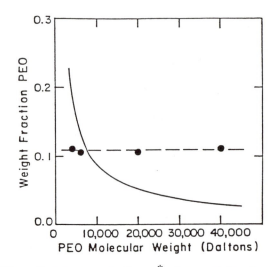

Figure 3. The polymer concentration, c^*, characterizing the transition from dilute to entangled (semidilute) polymer solution regimes, evaluated (see text) as a function of PEO molecular weight (full line), and the measured PEO concentration, c, in the top PEO-rich phase (filled circles).

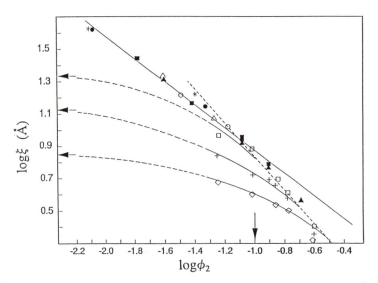

Figure 4. Logarithm of the static correlation length (in Angstroms), $\log\xi$, as a function of the logarithm of the PEO volume fraction, $\log\phi$, deduced from SANS measurements of PEO in D_2O. Polymer molecular weights in Da: (stars) 860,000; (open circles) 270,000; (open triangles) 160,000; (filled circles) 85,000; (filled squares) 45,000; (filled triangles) 21,000; (open squares) 9,000; (crosses) 4,000; (open diamonds) 1,500. The solid lines are interpolations of the experimental data and the dotted lines are extrapolations of the experimental data to a theoretical prediction for the size of an individual PEO coil (23).

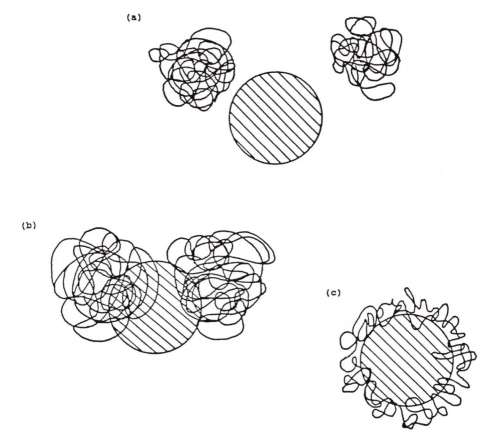

Figure 5. Three physical pictures representing the possible nature of the interactions between proteins and low molecular weight polymers: (a) Picture 1, physical exclusion only; (b) Picture 2, a weak attraction exists between the polymer and the protein in addition to physical exclusion; and (c) Picture 3, a stronger attraction between the polymer coils and the protein causes the formation of an adsorbed polymer layer about the protein.

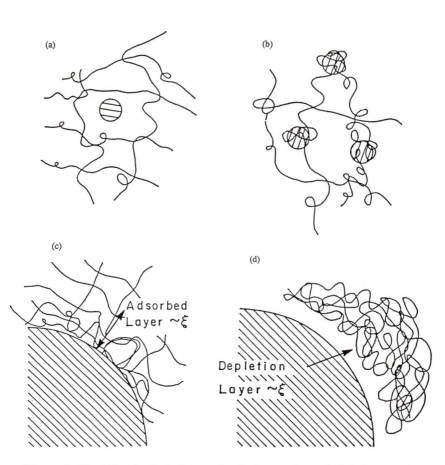

Figure 6. Possible physical pictures for the interactions of globular proteins and high molecular weight polymers: (a) very weak attraction or no attraction between the polymer mesh and the protein and $R_p<<\xi$; (b) strong attraction between the polymer mesh and the protein and $R_p<<\xi$; (c) strong attraction between the polymer mesh and the protein and $R_p>>\xi$; and (d) very weak attraction or no attraction between the polymer mesh and the protein and $R_p>>\xi$.

to the experimentally measurable protein partition coefficient, K_p, we were able to discriminate between the various physical pictures on the basis of the predicted (qualitative) influence of changes in PEO molecular weight (see Figure 1). Our findings can be summarized as follows. For Figures 5(a) and 5(b), the combined influence of repulsive steric (protein-polymer and polymer-polymer) interactions and weak attractive (protein-polymer) interactions on the change in the protein partition coefficient, $\Delta \ln K_p$, can be summarized by the equation *(21)*,

$$\Delta \ln K_p = -\Delta\left[\frac{N_2 U_p}{V}\left(1 + \frac{N_2 U_2}{2V} + O\left(\frac{N_2 U_2}{V}\right)^2\right) - \frac{k_1 \varepsilon a R_p^2}{N^{3/5}}\right] \qquad (1)$$

where N_2/V is the number density of polymer coils in solution, U_p is the excluded volume characterizing the steric interactions between a protein and a polymer coil, U_2 is the polymer-polymer effective excluded volume, ε is the local energy change (measured in units of kT) that accompanies the replacement of solvent at the protein surface by one polymer segment, N is the number of polymer segments per polymer coil (which is proportional to the polymer molecular weight, M_2), *a* is the size of a polymer segment, R_p is the radius of the protein, and k_1 is an order unity numerical prefactor. The first term in Eq.(1), $N_2 U_p/V$, describes the influence of repulsive steric interactions between proteins and polymers on $\Delta \ln K_p$, and results from a competition between N_2/V (which decreases with M_2, at a constant weight fraction of PEO, as $N_2/V \sim 1/M_2$) and U_p (which increases with M_2 as $U_p \sim M_2^\alpha$). It turns out that with increasing M_2, due to the increasing permeability of the polymer coils to the proteins, the exponent α is constrained within the range $0 < \alpha < 1$. Specifically, in the limit when the polymer coil is much larger than the protein, α tends to unity, which reflects the ability of the protein to penetrate the volume occupied, on average, by the polymer coil, and interact with the entire *length* of the polymer chain comprising the coil. Clearly, the combined influence of M_2 on N_2/V and U_p results in an increase (less negative) in $\Delta \ln K_p$ with increasing PEO molecular weight (see Eq.(1)), a trend which is opposite to that observed experimentally (see Figure 1).

In Eq.(1), the bracketed expansion in terms of $N_2 U_2/V$ describes the influence of polymer-polymer interactions on the protein partitioning behavior. Note that polymer-polymer interactions influence the protein chemical potential and thus affect the protein partitioning behavior. Similarly to the discussion presented above, the functional dependence of this term on M_2 can be expressed as $N_2 U_2/V \sim M_2^{\alpha-1}$, where now $\alpha = 9/5$ rather than 1, which results in an increase of this term with M_2. The reason for the rather different functional dependence of the polymer-polymer excluded volume on M_2, as compared to that of the protein-polymer excluded volume, is that the *two* polymer coils are of equal size at all PEO molecular weights, while the penetrability of a polymer coil to the protein increases with increasing PEO molecular weight. In summary, the influence of polymer-polymer steric interactions is to decrease $\Delta \ln K_p$ (increase the protein chemical potential) which is consistent with the experimentally observed influence of changes in PEO molecular weight on the protein partition coefficient (see Figure 1).

Although it is apparent from the first two terms in Eq.(1) that the predicted influence of PEO molecular weight on K_p results from a delicate balance reflecting opposing influences of the protein-polymer and polymer-polymer steric interactions, it is clear that in order to evaluate the additive influences of these interactions, a more quantitative approach is needed.

To meet this need, protein-polymer and polymer-polymer steric interaction potentials were evaluated *(22)* by combining a simple Monte-Carlo scheme (used to generate polymer-coil configurations in the vicinity of a globular protein molecule*(35)*) and the polymer solution theory of Flory and Krigbaum *(36)*. Through a recasting *(22)* of the interaction potentials in terms of effective hard-sphere potentials (which are non additive), the osmotic pressure and the chemical potentials of the colloid and the polymer were evaluated *(37,38)*. Although the treatment of steric interactions revealed the importance of including the deformability and penetrability of the polymer coil to the protein in determining the associated thermodynamic properties of solutions of proteins and polymers, it was not possible to predict the experimentally observed trends in the partitioning behavior of cytochrome-c, ovalbumin, bovine serum albumin, lactate dehydrogenase and catalase with increasing PEO molecular weight on the basis of repulsive steric interactions alone *(22)*.

In view of the fact that repulsive steric interactions alone could not account for the observed protein partitioning behavior, the last term in Eq.(1) was introduced *(21,22)* to examine the influence of a weak attractive interaction between the protein molecules and the PEO coils on the standard-state chemical potential of the protein. While the attractive interaction between a polymer coil and a protein molecule was shown to increase with increasing polymer molecular weight, when the opposing influence of the decreasing number density of polymer coils, N_2/V, was included, a decrease in the protein partition coefficient was predicted *(21)*. That is, the influence of weak attractive protein-polymer interactions is consistent with the trends observed experimentally in response to an increase in PEO molecular weight (see Figure 1). However, as stressed above, it is the additive effects of repulsive steric protein-polymer and polymer-polymer interactions, combined with weak attractive interactions between the polymers and the proteins, that determines the overall partitioning behavior of the protein. Using the equation-of-state/Monte Carlo approach (see above), we found *(22)* that, in order to account for the experimentally observed trends in the protein partition coefficient, the strength of the attractive interaction between a polymer-coil segment and a protein molecule had to be increased with protein radius, R_p, where $17\text{Å}<R_p<51\text{Å}$, from order 0.01kT to 0.1kT (per EO segment at the protein surface). However, the overall interaction (which includes the steric interaction) remained strongly repulsive. For example, for the protein ovalbumin ($R_p=29\text{Å}$), an attractive interaction energy of approximately 0.05kT (per EO segment at the protein surface) was shown to reproduce the experimentally observed protein partitioning behavior, and the protein-polymer second virial coefficient corresponding to this case was evaluated to be approximately 80% of the value predicted for the case of purely repulsive steric interactions *(22)*.

In addition, the presence of a strong attraction between the polymer coils and the proteins, and the associated formation of an adsorbed polymer layer at the surface of the proteins (see Figure 5(c)), are predicted to lead to a new partitioning behavior that has not yet been realized experimentally *(21)*.

In PEO solutions containing identifiable polymer coils, to further elucidate the influence of repulsive steric and attractive forces on the interactions of globular hydrophilic proteins and PEO, we performed SANS measurements of (1) aqueous solutions of PEO, (2) aqueous solutions of bovine serum albumin (BSA), and (3) aqueous solutions containing a mixture of BSA and PEO.

For PEO 8,650 Da, we found *(23)* that the measured excess scattering intensity, $I^{ex}(q)$, from an aqueous mixture of BSA and PEO is negative (see Figure 7), indicating that the net protein-polymer interactions are repulsive *(39)*. In order to provide a more quantitative interpretation of the SANs results, the excess scattering intensity was modeled *(23)* using effective hard-sphere potentials *(40-44)* for the protein-polymer interactions. Although the net interaction between the 8,650 Da PEO and BSA was found to be strongly repulsive, an effective protein size of 29Å was found to describe the excess scattering intensity. This value is significantly smaller than the 37Å expected for purely repulsive steric interactions *(22)*, thus suggesting the existence of an attractive interaction between a PEO coil and BSA. This conclusion is supported by an analysis which utilizes the Baxter sticky hard-sphere model *(45-47)*. Specifically, an attractive interaction energy of 0.05kT (τ=1.5 in Figure 7) (per polymer segment interacting with BSA) was necessary to describe the influence of PEO-BSA interactions on the SANS intensity measurements (see Figure 7).

Proteins in Entangled Polymer Solutions

In two-phase aqueous PEO-dextran systems containing high molecular weight polymers, the coexisting solution phases contain entangled webs of polymer, within which the identities of the individual polymer coils are lost. In such two-phase aqueous polymer systems, the partitioning of proteins is insensitive to the molecular weight of the polymers but is sensitive to their concentration and other factors (see Figures 1 and 2).

Because the partitioning behavior of proteins in two-phase aqueous polymer systems reflects the *relative interactions* between the proteins and the *two* coexisting polymer solution phases, and because the independent control of the polymer concentration in only one of the two coexisting phases is not possible, we have explored an alternative experimental technique, namely, the measurement of the partitioning of proteins between an entangled PEO solution phase and an aqueous (polymer-free) phase using a diffusion cell *(24)*.

In this case, the partition coefficient of the protein, K_p, is defined as the ratio of the protein concentrations in the top (PEO-free) and bottom (PEO-rich) compartments, respectively. Figure 8 presents the partition coefficients measured as a function of the concentration of PEO in the bottom compartment, ϕ, for cytochrome-c in buffered salt solutions having concentrations of 0.05M Na_2SO_4 and 0.1M NaCl, and for ribonuclease-a in buffered salt solutions having concentrations of 0.05M Na_2SO_4. The partition coefficients of both proteins are observed *(24)* to increase with PEO concentration, reflecting the tendency of the proteins to partition away from the bottom PEO-rich solution phase. The net repulsive interaction of the proteins with the PEO-rich phase is consistent with the observed interactions of the same proteins and PEO coils in two-phase aqueous PEO-dextran systems (see the discussions above and *(21-23)*).

In view of the potentially important effect of the Donnan equilibrium of charged species in similar systems, for example, two-phase aqueous polymer systems *(29,30)*, it is important to note that several observations indicate that this effect is not a dominant one in the diffusion cell experiments *(24)*. Specifically, (i) relatively small differences in the partitioning of cytochrome-c were observed to accompany the substitution of different salt types, and (ii) the PEO concentration in the PEO-rich compartment is an order of magnitude smaller than that encountered in typical two-phase aqueous polymer systems. As a result, the effective electrical potential difference across the membrane was estimated to be 0.2mV or less *(24)*.

To permit a comparison of the cell diffusion experimental results with scaling predictions of the protein partitioning behavior, the partition coefficients were correlated using the scaling form, $\ln K_p = A\phi^\gamma$, where ϕ is the volume fraction of PEO in the PEO-rich compartment *(24)*. The values of lnA and γ extracted from Figure 8, according to this correlation, are shown in Table I. Specifically, the exponent value, γ=1.22±0.06, was common to the three partitioning curves presented in Figure 8, whereas the prefactor A was observed to vary with the type of protein (but not with the type of salt). In order to interpret these experimental findings, we identified the relevant length

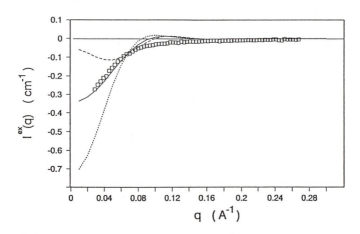

Figure 7. Excess neutron scattering intensity, $I^{ex}(q)$, as a function of q, the magnitude of the scattering vector, for a solution of 9.9 g/l BSA and 5.9% w/w PEO 8,650 Da in D_2O: (squares) experimental measurement; theoretical predictions using a sticky hard-sphere mixture structure factor and 3 different stickiness parameters, $[\tau_{p2}$ (dashed) 0.3, (solid) 1.5, (dotted) ∞].

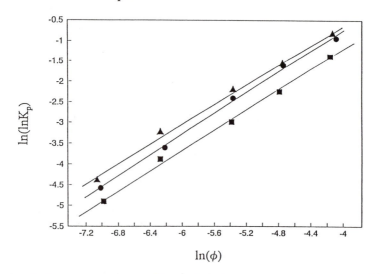

Figure 8. Protein partition coefficient, $\ln(\ln K_p)$, plotted as a function of PEO concentration in the bottom phase, $\ln\phi$: cytochrome-c in 0.05M sodium sulphate (triangles), cytochrome-c in 0.10M sodium chloride (circles), and ribonuclease-a in 0.05M sodium sulphate (squares). All solutions contained 10mM sodium phosphate (pH 7.0 buffer) and 1.5mM sodium azide. The molecular weight of the PEO was approximately 5 000 000 Da.

Table I. Exponents, γ, and prefactors, lnA, which relate the partitioning of cytochrome-c and ribonuclease-a into an entangled PEO solution in the presence of 0.1M NaCl and 0.05M Na_2SO_4

Protein	Salt	γ	lnA
cytochrome-c	0.1 M NaCl	1.20±0.04	4.17±0.09
cytochrome-c	0.05 M Na_2SO_4	1.26±0.03	4.29±0.07
ribonuclease-a	0.05 M Na_2SO_4	1.22±0.04	3.64±0.10

scales corresponding to the experimental situation under consideration. These are, the sizes of the proteins (cytochrome-c [R_p=17Å] and ribonuclease-a [R_p=16Å]), and the size of the polymer mesh (over the range of PEO concentrations 0.08% w/w to 1.7% w/w, the corresponding polymer mesh size is 30Å<ξ<300Å *(24)*). That is, the range of conditions explored were consistent with the constraint R_p<<ξ, or, at least, R_p<ξ.

In view of the magnitudes of these length scales, as well as the important role that a weak attractive interaction energy appears to play in determining the overall nature of the interactions between PEO *coils* and some globular proteins (in addition to repulsive steric interactions), we examined scaling predictions for the protein partition coefficient for the cases R_p<<ξ (see Figure 6(a)), and R_p>>ξ (see Figure 6(d)), in the presence of both repulsive steric and weak attractive interactions between the proteins and the polymer net. The results can be summarized as follows *(24)*:

and

$$\ln K_p \approx \phi \left(\frac{R_p}{a}\right)^{4/3}\left[1 - k_2\varepsilon\left(\frac{a}{R_p}\right)^{1/3}\right] \qquad a<<R_p<<\xi \qquad (2a)$$

$$\ln K_p \approx \phi^{9/4}\left(\frac{R_p}{a}\right)^{3}\left[1 - k_3\varepsilon\left(\frac{a}{R_p}\right)\right] \qquad a<<\xi<<R_p \qquad (2b)$$

where k_2 and k_3 are order unity numerical prefactors. From an inspection of Eqs.(2a) and (2b) it is evident that the experimentally deduced exponent value of γ=1.2 (see Figure 8) is close to the exponent value of γ=1 predicted in the limit R_p<<ξ (see Eq.(2a)), although the slightly higher than unity value is suggestive of a crossover towards the value of γ=9/4 predicted in the limit R_p>>ξ (see Eq.(2b)). However, in view of the similar sizes of the two proteins (see above), it is not possible to account for the different experimentally observed prefactors, A, (see Table 1) on the basis of size unless a different value of ε is assigned to each of the proteins *(24)*. Interestingly, the experimental observation that these proteins partition differently into entangled PEO solutions is consistent with the observed different partitioning behavior of these two proteins in two-phase aqueous polymer systems *(12)*.

Concluding Comments

We have reviewed our recent theoretical and experimental studies of the physical mechanisms responsible for protein partitioning in two-phase aqueous polymer systems *(21-25)*. It is gratifying to see that from a common physical basis, our theoretical approach, which combines polymer-scaling concepts, statistical-thermodynamic descriptions, and liquid-state theory, is capable of predicting both thermodynamic and structural properties of dilute and entangled polymer solutions containing globular proteins. The theoretical predictions, substantiated by independent experiments involving equilibrium protein partitioning studies and small-angle neutron scattering measurements, seem to indicate that protein partitioning in two-phase aqueous polymer systems reflects a delicate balance between repulsive steric protein-polymer and polymer-polymer interactions and weak attractive protein-polymer interactions. We hope that the philosophy behind our approach, as well as the detailed physical mechanisms which have been identified, will provide a solid foundation for future developments in the field of protein-polymer interactions with particular emphasis on protein separations.

Acknowledgments

Support for this work was provided by the National Science Foundation (NSF) through the Biotechnology Process Engineering Center at MIT under Grant CDR-88-03014, and a NSF Presidential Young Investigator (PYI) Award to DB, and by the Whitaker Foundation. In addition, DB was supported through a NSF Grant No. DMR-84-18718 administered by the Center for Materials Science and Engineering at MIT. DB is also grateful for the support by the Texaco-Mangelsdorf Career Development Professorship at MIT, as well as to BASF, British Petroleum America, Exxon, Kodak, and Unilever for providing PYI matching funds. NLA is grateful to the George Murray Scholarship Fund of the University of Adelaide, Australia for financial support.

References

(1) Tokiwa, F.; Tsujii, K. *Bull. Chem. Soc. Japan* **1973**, 46, 2684.
(2) Shirahama, K. *Colloid & Polymer Sci.* **1974**, 252, 978.
(3) Shirahama, K.; Ide, N. *J. Colloid Interface Sci.* **1976**, 54, 450.
(4) Cabane, B. *J. Phys. Chem.* **1977**, 81, 1639.
(5) Cabane, B.; Duplessix, R. *J. Physique* **1982**, 43, 1529.
(6) Cabane, B.; Duplessix, R. *J. Physique* **1987**, 48, 651.
(7) Goddard, E.D. *Colloids and Surfaces* **1986**a, 19, 255.
(8) Goddard, E.D. *Colloids and Surfaces* **1986**b, 19, 301.
(9) Heller, W.; Pugh, T.L. *J. Chem. Phys.* **1956**, 22, 17.
(10) Woodhead, J.L. *J. Physique Colloque C1* **1986**, 47, C1-3.
(11) Napper, D.H. *Polymeric Stabilization of Colloidal Dispersions*; Academic Press, London, **1983**.
(12) Albertsson, P.A. *Partition of Cell Particles and Macromolecules;* Wiley, New York, **1986**.
(13) Walter, H.; Brooks, D.E.; Fisher, D., Eds, *Partitioning in Aqueous Two-Phase Systems*; Academic Press, New York, **1985**.
(14) Kula, M.R.; Kroner, K.H.; Hustedt, H.; *Adv. Biochem. Eng.*, **1982**, 24, 73.
(15) Abbott, N.L.; Hatton, T.A. *Chem. Eng. Progress* **1988**, 84(8), 31.
(16) Cleland,. J.L., Massachusetts Institute of Technology, PhD Thesis, **1991**.

(17) Brooks. D.E.; Sharp, K.A.; Fisher, D. Chapter 2, *Partitioning in Aqueous Two-Phase Systems*, Eds Walter, H; Brooks, D.E.; Fisher, D. Academic Press, New York, 1985.
(18) Baskir, J.N.; Hatton, T.A.; Suter, U.W. *Biotechnol. Bioeng.*, **1989**a, 34, 541.
(19) Abbott, N.L.; Blankschtein, D.; Hatton, T.A. *Bioseparation*, **1990**, 1, 191.

(20) de Gennes, P.-G. *Scaling Concepts in Polymer Physics*; Cornell University Press, Ithaca and London, **1988**.
(21) Abbott, N.L.; Blankschtein, D.; Hatton, T.A. *Macromolecules*, **1991**, 24, 4334.
(22) Abbott, N.L.; Blankschtein, D.; Hatton, T.A. *Macromolecules*, **1992**, 25, 3917.
(23) Abbott, N.L.; Blankschtein, D.; Hatton, T.A. *Macromolecules*, **1992**, 25, 3932.
(24) Abbott, N.L.; Blankschtein, D.; Hatton, T.A. *Macromolecules*, **1992**, 25, 5192.
(25) Abbott, N.L.; Blankschtein, D.; Hatton, T.A. *Macromolecules*, **1992**, in press.
(26) Albertsson, P.-A.; Cajarville, A.; Brooks, D.E.; Tjerneld, F. *Biochim. Biophys. Acta* **1987**, 926, 87.
(27) Hustedt, H.; Kroner, K.H.; Stach, W.; Kula, M.-R. *Biotechnol. Bioeng.* **1978**, 20, 1989.
(28) Sasakawa, S.; Walter, H. *Biochemistry* **1972**, 14, 2760.
(29) Johansson, G. *Biochim. Biophys. Acta* **1970**, 221, 387.
(30) Johansson, G. *Molec. Cell Biochem.* **1974**, 4, 169.
(31) Albertsson, P.-A.; *Adv. Prot. Chem.* **1974**, 24, 309.
(32) Alexander, S.; *J. Physique* **1977**, 38, 977.
(33) Pincus, P.A.; Sandroff, C.J.; Witten, T.A. *J. Physique* **1984**, 45, 725.
(34) Marques, C.M.; Joanny, J.F. *J. Physique* **1988**, 49, 1103.
(35) Hermans, J. *J. Chem. Phys.* **1982**, 77, 2193.
(36) Flory P.J.; Krigbaum W.R. *J. Chem. Phys.* **1950**, 18, 1086.
(37) Carnahan, N.F.; Starling, K.E.; *J. Chem. Phys.*, **1969**, 51, 635.
(38) Carnahan, N.F.; Starling, K.E.; *J. Chem. Phys.*, **1970**, 53, 600.
(39) Cabane, B. in *Surfactant Science Series*, Volume 22, Ed., R. Zana, **1987**
(40) Ornstein, L.S.; Zernike, F., *Proc. K. Ned. Akad. Wet.*, **1914**, 17, 793.
(41) Percus, J.K.; Yevick, G.J., *Phys. Rev.*, **1958**, 110, 1.
(42) Lebowitz, J.L., *Phys. Rev.*, **1964**, 133, A895.
(43) Ashcroft, N.W.; Langreth, D.C., *Phys. Rev.*, **1967**, 156, 685.
(44) Ashcroft, N.W.; Lekner, J., *Phys. Rev.*, **1966**, 145, 83.
(45) Baxter, R.J., *J. Chem. Phys.*, **1968**, 49, 2770.
(46) Robertus, C.; Philipse, W.H.; Joosten, J.G.H.; Levine, Y.K., *J. Chem. Phys.*, **1989**, 90, 4482.
(47) Robertus, C.; Joosten, J.G.H.; Levine, Y.K. *J. Chem. Phys.*, **1990**, 93, 7293.

RECEIVED January 27, 1993

KINETICS AND CONFIGURATIONS OF ADSORBED CHAINS

Chapter 7

Kinetics of Grafted Chains in Polymer Brushes

Z. Gao and H. D. Ou-Yang

Department of Physics and National Science Foundation—
Industry–University Cooperative Research Center, Polymer Interfaces
Center, Lehigh University, Bethlehem, PA 18015

Dynamic Light Scattering technique was used to study the exchange
kinetics of chains between those in the brush and those in the
solution for polymers end-adsorbing onto colloidal surfaces. In the
experiment, the brush formed by long chains was allowed to fully
develop and then the measurements of the brush height were made
as a function of time starting when short chains were added into
solution. The brush height decreases because the short chains
invade into the brush and desorb part of the long chains. The
relaxation of the brush height can be characterized by a stretched
exponential function and the full equilibrium could take as much as
20 hours. The roles of long and short chains were also reversed for
checking the reversibility and for determining the equilibrium state.
This study reveals possibly a major expulsion mechanism for chains
anchored in a brush.

Strongly grafted polymers have many applications due to their unique ability to
form polymer brushes at the solid-solvent interface (1). The extended polymer
chains could provide effective colloidal stability and reduce shear friction between
close surfaces (2). Extensive studies, both experimentally and theoretically, have
yielded considerable fundamental understanding of the polymeric brushes (3).
For example, the relations between the brush thickness, grafting density, molecular
weight of the polymer and binding energy of the anchoring group as well as the
segment density profile in the brush are now fairly well understood. However, the
progress as in dynamics of polymer brushes, such as adsorption, desorption
kinetics, reversibility, chain exchange and desorption mechanism have emerged only

0097–6156/93/0532–0070$06.00/0

recently (*4*). Among them, Ligoure et al. gave a theoretical description of the adsorption and desorption kinetics (*5*). Experimentally, chain exchange in polymer brushes due to chain invasion have been reported recently by the authors (*6*). Further experimental evidence obtained by Klein et al. shows that replacement of long chains in a brush by the short ones could be fairly complete (*7*). Theoretical treatment using Kremer's rate theory on chain expulsion mechanism introduced by Halperin (*8*) was subsequently modified by Milner to explain competitive adsorption and desorption (*9*).

The chain dynamics during brush formation include roughly three stages. The initial stage of freely diffusive chains adsorb onto relatively bare surface. The second stage starts when chains must penetrate through a polymer layer of progressively increasing density. The final one is the brush completion stage where chains in the solution and in the brush are constantly exchanging but maintaining a macroscopic balance (*5*). Depending on the binding energy of adsorbing anchor and chain length, the kinetics of polymer adsorption can have a very large time span which makes measurements difficult. This is particular true in the last two stages.

The time span of the initial stage can usually be estimated by assuming the chain diffusion with the information of polymer concentration and inter-particle spacing (for adsorption onto colloidal particles). For example, assuming the chains stick to the surface at encounter, gaussian chains with a diffusion constant $D = 2 \times 10^{-7}$ cm^2/s, and the inter-particle spacing of 1 μm, the first stage would take about 0.1 second (too short for DLS measurements).

The build-up stage, where arriving chains have to penetrate through the not-quite-completed brush, takes a time which depends exponentially on the binding energy of the anchoring ends (*5*). For polymers with 1000 persistent length (Kuhn units) with 10 kT grafting energy the build-up stage takes a few minutes. However, this time can be as long as many hours for 30 kT of grafting anchor energy. During the brush growth process, but far from completion, the desorption of chains is negligible because the adsorption barrier is still much less than the desorption barrier.

The last stage is the only stage that the chain expulsion mechanism is playing a major role. The expulsion is important in the situations when selective adsorption, competitive adsorption are relevant. It is known that a fully developed polymer brush usually is non-washable. In other words, the desorption of chains is not possible simply by depleting free chains in solution; in this case the desorption is kinetically hindered (*10*). However, the free chains in solution can exchange with chains in the brush; i.e., the adsorption of chains facilitates other chains to desorb. Although the exchange flux near equilibrium can be negligibly small, when the system is brought away from equilibrium, the flux is detectable, which allows experimental investigation of the kinetics.

In this paper, we report results of experimental studies of the chain expulsion from fully developed brush triggered by invading chains. Preliminary results of this study were already reported recently (*11*), here we give a more detailed account of the investigation. The polymer exchange process occurred on the surface of polystyrene latex spheres. The thickness of fully developed polymer brushes on the latex sphere surface was observed to decrease when shorter polymer chains were sent to invade into the brush. Through the chain exchange kinetics study we could also gain useful information for times need for brush construction as well as some insights on the issues of reversibility and equilibrium of polymers in a brush. For technical reason, we used chains which can graft onto the surfaces with both ends. This might not be the easiest system for theorists, however, it added some twists to the experiment.

In order to assert the construction time of C16-100 brush, one could measure directly the growth of the brush height as a function of time. However, because the equilibrium density of the brush layer cannot be determined by DLS directly, even if we measured the brush height we still did not know if the brush density had reached the equilibrium profile. As a matter of fact, we did measure the growth of the brush height as a function of time, however, the brush height increased rapidly to its full value in less than 3 minutes, but we could not be sure that full density of the brush had achieved equilibrium in that time. An alternative approach to study the construction time is to examine the relaxation behavior in responding to a perturbation to the system at different incubation stage. We expect the system will response differently to the same perturbation if the system has not reached equilibrium. Once the relaxation behavior no longer depends on the incubation time, then this time can be thought of as the upper limit of the construction time. This second approach was tested and was used to compare with the first one.

Experimental

The Dynamic Light Scattering (DLS) technique was used to measure radii of the PS latex spheres with and without adsorbed polymer brushes. We could then deduce the polymer brush hydrodynamic layer thickness by taking the difference of the radii. DLS measures the intensity autocorrelation as a function of delay time, which gives information on the diffusion constant of particles in a dilute solution. The translational diffusion coefficient, D_t, is related to the solution temperature T, particle radius r, and solvent viscosity η by the Stokes-Einstein relation:

$$D_t = \frac{kT}{6\pi\eta r} \tag{1}$$

By knowing T and η, we can determine the particle radius. The polymer layer thickness, h, can then be readily determined by the following relation:

$$h(t) = r_h(t) - r_o \tag{2}$$

where r_h and r_o are respectively the hydrodynamic radii of the particles with and without adsorbed polymers.

The polymers used in this study consist of water-soluble poly(ethylene oxide), or PEO, backbones terminated on both ends by hydrocarbon hydrophobes. The persistent length of PEO, consists of two monomer chemical units, is 4.2 Å long. Two different hydrophobic ends are used: $C_{16}H_{33}$ and $C_{20}H_{41}$. For simplicity, we use the symbol Cn-W to represent our sample polymers in which n represents the carbon number of the hydrophobe and W stands for the polymer molecular weight in thousands. For example, C16-100 stands for 100,000 g/mole PEO chains terminated by $C_{16}H_{33}$ on both ends. In aqueous solution, the hydrophobic ends adsorb strongly onto the polystyrene (PS) latex particle surface whereas the backbones also adsorb but relative weakly. The bare PS particles have a diameter of 91±1 nm as measured by DLS. At sufficient concentrations, these polymers form a brush on the latex surface (*12*)(*13*).

To avoid both the particle interactions and the possible multiple light scattering, all the measurements were made at the 10^{-4} volume fraction of PS spheres in water. Since we fixed the PS particle concentration for the convenience of discussion, polymer concentrations are labeled in number of chains per sphere for simplicity. The solution concentration of the C16-100 polymers at 1000 chains per sphere and at the above volume fraction is about 4.2×10^{-5} g/ml. When fully saturated, the polymer brush has a surface coverage of 4.8 mg/m² (*14*). The fully developed layer thickness of brush formed by C16-100 polymer is 50 nm. Based on a simple assumption of uniform monomer distribution in the brush, the polymer concentration in the brush can be calculated as 0.096 g/ml. This concentration is 4 times C^*, where C^* the overlapping concentration of polymer chains in good solvents (*15*). The measurements are made at a fixed scattering angle of 135° and at a temperature of 23±0.1°C. To follow the kinetic process, the digital correlator was programmed to accumulate and record the scattering data every 4 minutes for the duration that the exchange process was taking place.

Results and Discussion

The first part of the experiments was to use the short invader chains to displace the polymer chains from a brush. Desorption of chains from fully developed brush by added invading chains is shown in figure 1. By "fully developed brush" we mean a brush that was allowed to develop for a time much longer than the construction time needed to form a brush (which, as shall be seen later, was typically less than one hour). The brush was formed by letting 8.4×10^{-5} g/ml C16-100 polymers co-exit with 91 nm PS spheres at 10^{-4} volume fraction for more than 48 hours. At this mixture, the number of chains per sphere added in solution is 2000. (We know from other experiments that some chains were present in free solution after the brush was formed.) In figure 1, the time zero is marked by the time at which the invading C16-17 chains were added to the mixture. The decay of the layer thickness indicates that part of the long chains in the brush have been replaced by the short chains; the degree of the replacement depended on the amount of the added short chains. Since the measurements were made at very low polymer and PS sphere concentrations, osmotic compression exerted on the brush by free polymers in solution was not important. Also note that the asymptotic values of the

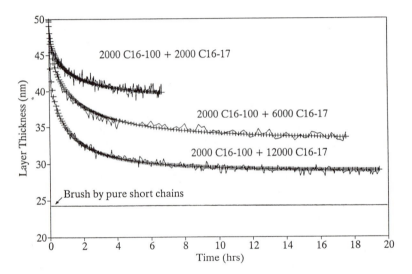

figure 1. Desorption of polymer chains from fully developed brush by adding invaders. Crossed lines are fitted to stretched exponential functions.

layer thickness after decay were still higher than the thickness of the brush formed by pure C16-17 chains, indicating that the replacement of long chains by the shorts were not complete. This was in contrast to the observation made by Klein that all of long chains were replaced by the shorts (7). We believe the difference between the two observation was due to the facts that grafted polymers in Klein's experiment had surface repelling backbones, ie., they were surfactant like polymers, whereas in this study the grafting polymers had weakly adsorbing backbones. It is known that for homopolymer adsorption, long chains dominate the surface, while for grafting surfactant polymers the short chains dominate the surface. For grafting polymers with weak adsorbing backbones we expect that both long and short chains stay on the surface as a result of a balance between end and backbone adsorption.

Stretched Exponential Behavior. The solid lines in figure 1 were actual fittings of the data to the stretched exponential function:

$$h(t) - h(\infty) = [h(0) - h(\infty)] \cdot \exp[-(\tfrac{t}{\tau})^{\alpha}] \qquad (3)$$

Surprisingly, we found that for all three curves the best fit exponents α were very close to 0.50. Although we did not expect the data to be fitted by a single exponential we did not expect all of the curves to be fitted by the same stretch exponent α either. As a matter of fact, as shown in Table 1, for all the measurements of C16-100 brushes invaded by C16-17 the exponents α falling in a very narrow range of 0.50 ± 0.02. In the group of C16-100/C16-34 the α were in the range 0.40 ± 0.03.

There are various ways to argue the significance of the stretch exponential relaxation behavior and the meaning of the exponents α depends on the exact physical process such as the so called parallel process or the correlated sequential process which are commonly used to explain glassy behavior. But there were other possibilities, here we have tried to rule out some. One question was that the system might have been driven too far away from the equilibrium. It could be argued that since systems are far away from equilibrium, during the relaxation process many time scales can involve and therefore yield a stretched exponential behavior. The distance from equilibrium will then determine the stretchness α of the exponential relaxation (Halperin, A., personal communication). In other words, α should depend on the invading polymer concentration; however, data in Table 1 showed otherwise. The other possible cause was the invader polymer polydispersity. Since we did not have the option to try monodisperse polymer samples we made the system more polydisperse by mixing C16-17 and C16-34 at equal molar ratio, but we found that the exponent α essentially unchanged. We have not checked the effect of polydispersity of chains in the brush. A theoretical treatment assuming distant dependent exchange probability can yield a stretching exponential behavior and will be published later (16).

The characteristic relaxation time, τ, although varying from system to system, was found to be on the order of one hour for all samples. Note that

TABLE 1. Values of α, τ and h(∞) in equation 3 for various samples

Surface Polymers (per sphere)	Invading Polymers (per sphere)	Relaxation Time τ (hrs)	α	Equilibrium Layer Thickness h(∞)(nm)
1000 C16-100	1000 C16-17	0.4	0.53	41.7
1000 C16-100	3000 C16-17	0.93	0.5	33.3
1000 C16-100	6000 C16-17	0.53	0.48	29.1
2000 C16-100	2000 C16-17	1.0	0.52	39.1
2000 C16-100	6000 C16-17	1.3	0.52	33.2
2000 C16-100	12000 C16-17	0.66	0.47	28.9
1000 C16-100	1500 C16-17 & 1500 C16-34	0.6	0.48	34.4
1000 C16-100	1000 C16-34	0.17	0.39	43.6
1000 C16-100	3000 C16-34	0.47	0.39	34.3
1000 C16-100	6000 C16-34	0.49	0.42	33
2000 C16-100	2000 C16-34	0.86	0.43	38.4
2000 C16-100	6000 C16-34	0.87	0.41	34.6
2000 C16-100	12000 C16-34	0.84	0.36	33.6
1000 C20-116	3000 C16-17	1.9	0.61	38.5
1000 C20-116	6000 C16-17	2.5	0.62	34.1
1000 C20-116	3000 C16-34	2.3	0.8	38.3

Note: Data taken for all exchange processes.

stretched exponential functions decay faster than the single exponential function in the times less than τ and much slower thereafter. It implies, in these measurements, most of the exchange activities happened in the first hour, whereas detailed adjustment of chain distribution could continue for twenty hours. The characteristic time τ varied with both the polymer concentration and the polymer species, but we have not found a systematic trend for them. For example, the τ for the group of C16-100 washed by C16-17 was not a monotonic function of invading-chain concentration. For the group of C16-100 washed by C16-34, we found that when the concentration of surface polymers was low, τ was monotonically increasing with the concentration of invading polymers; when the concentration of surface polymers was high, τ became independent of invading-polymer concentration.

Construction and Washing Times of the Polymer Brush. We stated earlier that the "fully developed brush was formed within about one hour"; how did we determine the brush construction time and how is it compared with the theory? We explain as follows. As is mentioned in the introduction, the only information we got from direct measurement of brush height growth as a function of time was that the hydrodynamic brush thickness was established in two or three minutes. From that alone we could not assess the full details of brush development. To do more, we took advantage of the chain exchange process by probing the system during the brush developing period. The idea was to invade the "polymer brush" by short polymer chains at different stage of brush incubation and observe the desorption process as a function of the incubation time. The brush incubation time at which substantial change of the desorption process occurs should mark the construction time. For this approach to work, it is important to realize that in the exchange process the relaxation of the brush height is described by a stretched exponential function characterized by two parameters, the exponent α and characteristic time τ. A "substantial change" in desorption process means a change in either α or τ. As shown in figure 2, the experimental results show that α and τ do not change appreciably over the time range from 15 minutes to more than 100 hours. Experimentally, there is no evidence that suggests an expectation of a construction time on the order of hundred hours or longer. But other direct measurements of brush growth suggested the construction process to be on the order of minutes. Combining these two measurements, we conclude that the brush construction time was less than 15 minutes.

Ligoure et al. have calculated the required construction time for single-end-grafted polymers with a surface repelling backbone. In their calculation the construction time τ_c is expressed by the following equation:

$$\tau_c = \frac{2}{3}\tau_2 \frac{e^{\gamma N \sigma_{eq}^{\frac{2}{3}}}}{\gamma N \sigma_{eq}} = \frac{8}{9}\left(\frac{3}{\pi^5}\right)^{\frac{1}{3}} \frac{a^5 N \eta_0}{\upsilon^{\frac{2}{3}}\phi_0} e^{\gamma N \sigma_{eq}^{\frac{2}{3}}} \qquad (4)$$

with $\tau_2 = \dfrac{N^2 \sigma_{eq} a^3 2^{\frac{1}{3}} \eta_0}{\pi kT \phi_0}$, $\gamma = \dfrac{3}{2} \left(\dfrac{\pi^2}{12} \right)^{\frac{1}{3}} \left(\dfrac{\upsilon}{a^3} \right)^{\frac{2}{3}}$ and $\sigma_{eq} = a^2/\Sigma$

where N is the number of persistent (Kuhn) length, a the Kuhn length, Σ the average surface area per chain, υ the excluded volume, η_0 the solvent viscosity and ϕ_0 the monomer volume fraction in the reservoir. Notice that in equation 3 the construction time goes exponentially to the deformation energy of a chain in a brush. This deformation energy is balanced by the binding energy Δ of grafting end and the entropic cost of a chain in the brush. As an approximation, we can ignore the entropic terms when the binding is strong ($\Delta >> kT$) and the solution concentration is not too low, therefore the exponent goes like Δ. Thus, a small change in the binding energy will change the time-scale significantly.

In the brush regime the above calculation could be applicable to the double-end-grafted polymers as an approximation as long as the backbone is long enough that the correlation between ends is negligible. Given the following values: a=0.42 nm, N=1136, υ=0.008 nm^3, k=1.38x10^{-16} erg/K, T=296 K, η_0=0.09325 poise, Σ=34.7 nm^2 (14) and ϕ_0=10^{-5} we estimated the construction time for C16-100 polymers to be about one minute, consistent with our measurements. The relatively short construction time in this study is quite different from the long time-scale of hours involved in homopolymer adsorption, for which many segments of the chain have to interact with the surface and rearrange themselves such that the free energy of the chain can be minimized (17). Since our brush contained mostly loops with small fractions of backbones on the surface, the rearrangement of backbones did not appear to play a major role during the construction of brush. In other words the construction time of a brush formed by polymer loops behaved like conventional brushes.

Ligoure et al. have also calculated the washing time of polymer brushes on the surface by pure solvent, which is given by the following equation:

$$\tau_w = \dfrac{3}{2} \gamma N \sigma_{eq} \tau_c \tag{5}$$

Substituting the above experimental values directly into the equation, we would get a washing time of about 3 minutes, too short comparing with observation where we found washing took longer than 48 hours (12). The fact that most of the chains in the brush form loops instead of single-ended grafting makes a difference here. Since there are two adsorbing ends on one chain, in order to wash off a chain, both ends must desorb from the surface before any readsorption can occur. Thus, the energy associated with desorption for polymer loops will be at least twice as large as the polymer with one end grafted. In addition, because of possible weak adsorption of a fraction of backbone segments, the binding strength of the polymer could be slightly larger. With this consideration we could push the estimated washing time to about 30 hours, much closer to the observation.

Enhanced Desorption of Long Chains by Introducing Short Chains. The explanation for the fact that it takes longer time for a polymer brush to be "washed off" from the interface by pure solvent than by other polymers or surfactants was not clear until the present experiment. The key issue is hinged upon the question of how a desorbed chain leaves the surface (how does a chain know when to leave). As usual, washing by solvent, thermal motion is sufficient for some grafted ends to leave the surface but this is only the beginning of the desorption process. For a chain to leave the brush completely, it has to reptate out of the brush (the monomer density in a brush is typically above C*) during which a reattachment of the anchor might happen if the driving force in the outgoing direction is weak. According to Halperin and Milner, this force can be deduced from the work needed to pull a chain into the brush from solution W(z)

$$W(z) = \frac{2f}{\pi} \left[\cos^{-1}\left(\frac{z}{h}\right) - \frac{z}{h}\left(1 - \frac{z^2}{h^2}\right)^{1/2} \right] \quad (6)$$

experienced by a chain in a brush, where f is the deformation energy of a chain in the brush, z the distance of the anchor from the surface and h the brush thickness (*9*). The outward driving force, or the outgoing polymer flux is proportional to the derivative of W(z) with respect to z. It is found that W'(z) is typically small. However, shorter invading chains see a much lower energy barrier than the long chains do because of less stretching in the brush, and therefore, can readily penetrate the brush before the long chains have a chance to escape. This creates a transient overcrowded state of chains in the brush. Crowding by the short chains create a higher local chain tension, ie., a steeper outward W'(z), making long chains easier to desorb (*9*). In figure 3, we show during the transient, when $\tau \geq 0$, the shadowed area in W(z) produced by the overcrowding near the surface increases the outgoing driving force W'(z). The transient time continues until the overcrowding is relieved by desorbing long chains. At long times, when $t >> \tau$, the W(z) assumes a new shape corresponding to a lower brush thickness at equilibrium. According the calculation, the enhancement factor for the long chain desorption rate is $(N_l/N_s)^{0.5}$ of the single-species washing rate, where N_l and N_s are polymerization indices of long and short chains, respectively.

Overshoot in Brush Height at Short Time. The above picture predicted that the brush height could overshoot its initial value immediately after the short chains invade into brush. Indeed, this was observed in our experiment (figure 4) when pre-adsorbed polymer brush was formed by C20-116. Four minutes (the time when the first data point was taken) after the invaders C16-34 were added, the brush height was 5 nm more than the original brush. The overcrowding persisted for at least 12 minutes before a normal desorption relaxation process took place. The data taken for $t > 30$ minutes was fitted to a stretched exponential function with $\alpha = 0.8$ and $\tau = 2.3$ hours.

The overcrowding configuration of the polymer brush during the transient,

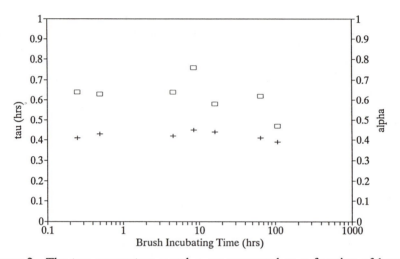

figure 2. The two parameters τ and α are measured as a function of brush incubation time.

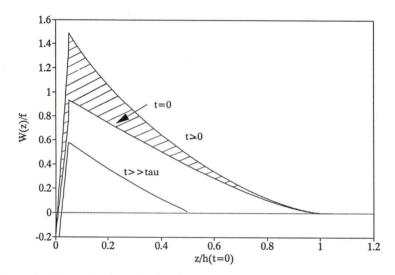

figure 3. Normalized work $W(z)/f$ as a function of distance z away from the adsorbing wall. $h(t=0)$ is the brush height before invaders were sent.

when short and long chains coexist, can be described as follows. The brush consists of two layers: the first layer closer to the surface is formed by strongly-stretched polymer segments belong to both the short and long chains, beyond this layer is layer containing mostly segments of long chains which are much less stretched.

Note that our observation of brush height overshoot was resolution limited. We don't know if the brush height was even higher during the first four minutes. We believe, however, that the transient overshooting stage will be prolonged if the brush is denser and thicker so that we could explain why overshooting was not observed when the less dense C16-100 brush was invaded. The brush formed by C16-100 was less dense and the transient time of overshooting was too short to be resolved by our measurement.

An additional contribution to the brush height overshooting could be that, initially, the short chains were only partially penetrating the brush and the dangling ends of short chains added to the original brush height. This was certainly plausible and we cannot rule out its possibility at this stage.

The Milner theory provides an intuitive picture of why invading chains facilitate the desorption of pre-adsorbed long chains, however we have some difficulties in using the theory to quantitatively explain the experimental observations. First of all the stretched (experimental) vs. single exponential (theoretical) behaviors, which, we feel, could be due to reasons intrinsic to the experiments such as two anchors per chain and backbones attractive instead of repulsive from the surface. Secondly, the theory predicts that the desorption rate C16-100 brush invaded by C16-17 would be shorter than that by C16-34 by a factor $[(N_l/N_s)^{0.5} = (34/17)^{0.5} = 1.41]$, however experimentally, they showed the same desorption rate. More importantly, the picture cannot be used to explain the case when invading chains are longer than the brush chains. In a brush form by surfactant polymers the long chain will not displace short chains in brush. But because of the weak backbone adsorption of our polymers, long and short chains co-exist in the final brush. To see this we reverse the adsorption order of long and short polymers, that is, we let short chains pre-adsorb on the surface and then introduce long chains.

Reversibility and Equilibrium. In figure 5, we show two curves the top one is the usual long chain brush invaded by short chains and the bottom one was made by letting the short C16-17 chains pre-adsorb on the surface and invade with the long C16-100 chains. The two curves represent systems with the same polymer composition with different kinetic process. At long times, both curve reach the same layer thickness with the final state obviously composing both long and short species. This procedure demonstrates that the system can reach equilibrium and the adsorption-desorption processes are reversible. Also we note that, in figure 5, the top process is slower than the bottom one. This would indicate that the exchange kinetics are governed by desorption and the top process was slow because it involved long chains escaping from a thicker layer. It will be interesting to see when equilibrium cannot be reached when the binding energy of ends is so strong

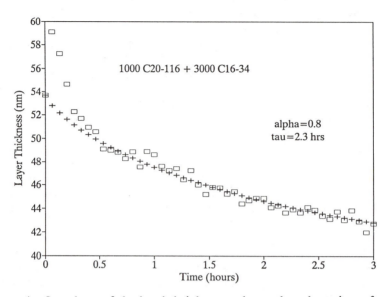

figure 4. Overshoot of the brush height was observed at short time after the brush formed by C20-116 was invaded by C16-34. Empty squares are experimental data and the crosses are fitted to the data after the overshoot.

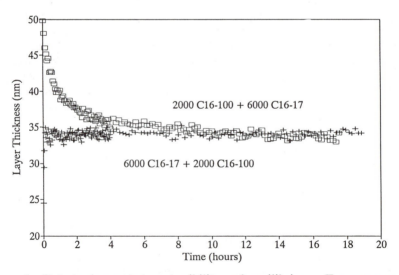

figure 5. Data to demonstrate reversibility and equilibrium. Top curve was taken when brush formed by C16-100 was invaded by C16-17 chains. The lower one was for the reverse process. (Reproduced with permission from reference 11. Copyright 1992 by Materials Research Society.)

that ends cannot desorb from the surface. In that case, the two curves in this figure 5 approach different values even at very long times.

Conclusions

Starting out trying to solve the puzzle that polymers are more effective than pure solvents in displacing chains from polymer brushes we have acquired a better understanding of polymer expulsion mechanism. The experiments we used here to study chain exchanging is probably one of the first direct measurements on kinetics of polymers in a brush. Local chain crowding, suggested by Milner, caused by invading chains provided sufficient driving force for chain desorption. Transient overshoot of brush height at the beginning of the exchange process further support the chain crowding picture. Good agreements were also reached between experiment and theory on the brush construction time, washing time by pure solvent. In the latter case, the fact that both ends of the polymer chain were anchoring has to be taken into consideration.

However, several key issues remain unsettled. The long chains replace short chains appeared even more effective than the reversed case. If this were not caused by the experimental artifact that the backbones adsorb, it could pose a challenge to the chain crowding picture which would predict that long chains could not displace the shorts. The stretched exponential relaxation during the exchange process, although bear a strong resemblance to that found in adsorbed polymers (6), is not expected in grafted polymer systems. Finally, the systems during adsorption and desorption have reached equilibrium and shown that the exchange process were reversible. In the future, it will be interesting to study kinetics of polymer exchange in systems that are prevented from reaching equilibrium due to kinetic hinderance.

Acknowledgments

We thank Union Carbide Chemicals for providing the polymer samples and D. Bassett and R. Jenkins for sharing detailed information about these polymers. We are grateful to S. Milner, A. Halperin and M. Daoud for their valued comments. This work is partly supported by a grant from American Chemical Society through The Petroleum Research Fund ACS-PRF# 25250-AC and a grant from National Science Foundation through the NSF/IUCRC Polymer Interfaces Center at Lehigh.

Literature Cited

1. Napper, D. *Polymeric Stabilization of Colloidal Dispersion*, (Academic: London) 1983.
2. Klein, J.; Perahia, D.; Warburg, S. *Nature*, **1991**, *352*, 143.
3. For a recent review, see Milner, S.T. *Science*, **1991**, *251*, 905, and references therein.

4. Leermakers, F.A.M.; Gast, A. *Macromolecules*, **1991**, *24*, 718;
   Frantz, P.; Granick, S. *Physical Review Letters*, **1991**, *66*, 899.
5. Ligoure, C; Leibler, L. *J. Phys. France*, **1990**, *51*, 1313.
6. Johnson, H.E.; Granick, S. *Science*, **1992**, *255*, 966.
7. Klein, J.; Kamiyama, Y.; Yoshizawa, H.; Israelachvili, J.N.; Fetters, L.J.;
Pincus, P. *Macromolecules*, **1992**, *25*, 2062.
8. Halperin, A. *Europhysics Letters*, **1989**, *8 (4)*, 351.
9. Milner, S. *Macromolecules*, **1992**, *25*, 5487.
10. de Gennes, P.G. *Advances in Colloid and Interface Science*, **1987**, *27*, 189.
11. Gao, Z.; Ou-Yang, H.D. IN *Complex Fluids*; Sirota, E.B.; Weitz, D.; Witten,
Tom; Israelachvili, J, Eds; MRS Symposium Proceedings; Publisher: MRS,
Pittsburgh, PA, 1992, Vol. 248; pp 425-430.
12. Ou-Yang, H.D.; Gao, Z. *J. Phys. II (France)*, **1991**, *1*, 1375.
13. Alexander, S. *J. de Physique*, **1977**, *38*, 983.
14. Jenkins, R.D; Ph.D thesis, Lehigh University, PA, U.S.A. (1990).
15. de Gennes, P.G. *Scaling Concepts in Polymer Physics*, Cornell University
Press.
16. Vugmeister, B; Hong, D.C.; Ou-Yang, H.D. and Gao, Z.; to be published.
17. de Gennes, P.G. *Journal de Physique*, **1976**, *37*, 1445.

RECEIVED December 22, 1992

Chapter 8

Configuration of Polyelectrolytes Adsorbed onto the Surface of Microcapsules

Etsuo Kokufuta

Institute of Applied Biochemistry, University of Tsukuba, Tsukuba, Ibaraki 305, Japan

Configurational changes of polyelectrolytes adsorbed onto a hydrophobic surface were studied as a function of pH by examining the inward permeation of n-propyl alcohol through poly(styrene) microcapsules with an outer surface layer onto which the polyelectrolytes had been adsorbed. Three polyelectrolytes were used: alternative copolymers of maleic acid with styrene, copoly(MA, St), and with methyl vinyl ether, copoly(MA, MVE), and poly(iminoethylene), PIE. It was found that both adsorbed copoly(MA, St) and PIE undergo rapid configurational changes at the pH levels which cause their conformational transitions in aqueous solution as elucidated by electrophoretic and viscometric measurements. Moreover, the configuration of the copoly(MA, MVE) was found to vary gradually depending on the pH of the outer media in a way which was comparable to its conformational changes in the aqueous solution. When enzymes such as invertase were loaded into microcapsules with a surface layer of PIE or copoly(MA, St), the pH-sensitive on/off control of enzyme reactions could be successfully performed through drastic alterations in the substrate permeability of the capsules brought about by pH-conditioned configurational changes in the adsorbed polyions.

Polyelectrolytes adsorb onto solid surfaces from their aqueous solutions in either flat or looped form (for example, refs. 1- 4). Adsorbed polyions are expected to undergo configurational changes on surfaces when such external factors as pH and ionic strength are altered. However, in contrast to the conformational changes of polyions in aqueous media which can be examined using various experimental methods, the configurational changes occurring on solid surfaces are very difficult to monitor; thus, few studies have dealt with this subject to date.

The present study has adopted a promising approach for investigating pH-induced configurational changes in polyions adsorbed onto the surface of a hydrophobic membrane. The idea used here is based on the examination of the inward permeation of small water-soluble molecules through a semipermeable capsule membrane with a layer of adsorbed polyion on its outer surface. A change in permeation is expected when the configuration of adsorbed polyions varies from looped to flat (or vice versa) as illustrated in Figure 1.

0097−6156/93/0532−0085$06.00/0

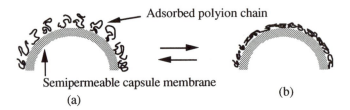

Figure 1. Schematic representation of configurational change of polyions adsorbed onto outer surface of microcapsule: (a) looped form; (b) flat form.

The regulation of such a configurational change using externally applied stimuli (such as altering the pH level) is expected, on the other hand, to provide a useful tool for controlling the inward or outward permeation of solutes through a capsule membrane. The present study also attempts the encapsulation of an enzyme in microcapsules and the on/off control of the enzymic process through the regulation of the configurational changes of the polyions adsorbed onto the outer surface of the microcapsules *(5,6)*.

Polyelectrolyte Samples and Their Conformational Changes in Aqueous Solutions

The primary requirement in the present case is to elucidate pH-induced conformational changes of polyelectrolytes in their aqueous solutions. This would provide a way to discuss changes in the adsorbed polyion configuration through comparison with the pH dependence of the permeability of capsule membranes with an adsorbed polyion layer. The conformation of polyions varies depending on the balance between the repulsive and attractive forces along the polymer chains. A repulsive force is usually electrostatic in nature and controlled by pH and ionic strength. When a repulsive force overcomes an attractive force such as hydrogen bonding or hydrophobic interaction, the polyion chain should extend discontinuously in some cases or continuously in others, as is seen in the volume changes of polymer gels *(7)*. Thus, polyelectrolytes, the conformational changes in which can be explained using such concepts, should provide good samples for the present purpose.

Materials and Methods. The following polyelectrolytes were used: copolymers of maleic acid (MA) with methyl vinyl ether (MVE), copoly(MA, MVE), and with styrene (St), copoly(MA, St), and poly(iminoethylene) (PEI, branching type).

$$+CH-CH_2-CH\!\!-\!\!-CH\}_n \qquad ---(CH_2-CH_2-N\}_y---+CH_2\text{-}CH_2-NH_2)_z$$

OCH₃   COOH COOH

Copoly(MA, MVE)
Mn = 4.16×10⁵

$$+CH-CH_2-CH\!\!-\!\!-CH\}_m$$

COOH COOH

Copoly(MA, St)
Mn = 2.60×10⁵

PIE
Mw = 1.1×10⁵
$x : y : z$ = 2:1:1

Viscometric and electrophoretic measurements were carried out to study the conformational changes of the above in aqueous solutions. The polyelectrolytes were dissolved in buffer solutions (acetate, pH < 6; phosphate, pH 6 ~ 8; carbonate, pH > 8) adjusted to ionic strength 0.1, and then dialyzed against the same buffer solution in a cellophane tube until Donnan equilibrium was reached. The sample solutions were prepared at different concentrations by diluting with the buffer against which the polymer solution had been dialyzed. Electrophoretic mobility was measured at 25±0.01 °C with a Hitachi Tiselius Electrophoresis apparatus (model HTD-1). A microtype cell with a cross-sectional area of 0.19 cm^2 was used. The viscosity was measured at 25±0.005 °C using an Ubbelohde viscometer with a follow time of 305 s for water at 25 °C. Both measurements were done over a polymer concentration range of 0.005 to 0.2 g/dL, and the linear plots of the mobility or viscosity against the concentration were extrapolated to zero concentration as described previously *(8,9)*.

Results and Discussion. Figure 2 shows the pH dependence of the limiting mobility ($U{\to}0$) and the intrinsic viscosity ($[\eta]$) for three polyelectrolytes. The mobility curves of copoly(MA, MVE) and copoly(MA, St) display an increase over pH 3 to 5, a plateau near pH 5 ~ 6, and a decrease over the alkaline pH range. In previous studies *(8,9)* of both copolymers by means of potentiometric titration, it was found that two COOH groups in an MA unit dissociate independently in two stages below and above pH 5 ~ 6:

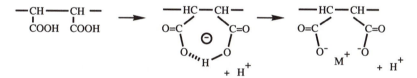

The variations in the mobility curves which appeared in the pH regions below and above the plateau can thus be assigned to the first and second dissociation stages of the copolymer-bound COOH groups, respectively. As a result, the copolymer ions are found to carry a large portion of the negative charges in the first dissociation stage. A slight decrease in the polyion charges (pH > 7 ~ 8), as revealed by lowering the mobility in the second dissociation stage, seems to be attributable to a possible trapping of a counter-ion between the two COO$^-$ ions in the MA unit.

These alterations in the charges of copolymer ions directly affect their conformations through strong electrostatic interactions, since the viscosity curves display a resemblance to the mobility curves; that is, an increase in the polyion charges as a result of the dissociation of COOH leads to the expansion of the polyion chain as indicated by a rise in the viscosity with increasing pH. However, a detailed comparison of the viscosity curves of copoly(MA, MVE) and copoly(MA, St) shows that the viscosity change of the latter copolymer rapidly occurs within a very narrow range near pH 4.5. Several previous studies of copoly(MA, St) *(8,10,11)* have demonstrated that its conformation is affected not only by an electrical repulsion force between the COO$^-$ ions, but also by a hydrophobic interaction (as an attractive force) between the phenyl groups. When the hydrophobic force is overcome by the electrical force, a conformational transition from a "tightly coiled chain" to an extended one occurs as illustrated in Figure 3. Therefore, the pH-induced change in the conformation of copoly(MA, St) is discontinuous, while copoly(MA, MVE) undergoes a continuous conformational change with pH.

A conformational transition was also demonstrated by the mobility and viscosity curves for PIE. A rapid increase in the viscosity with decreasing pH could indicate the conformational transition of the polyion from a tightly coiled chain (pH > 6) to a

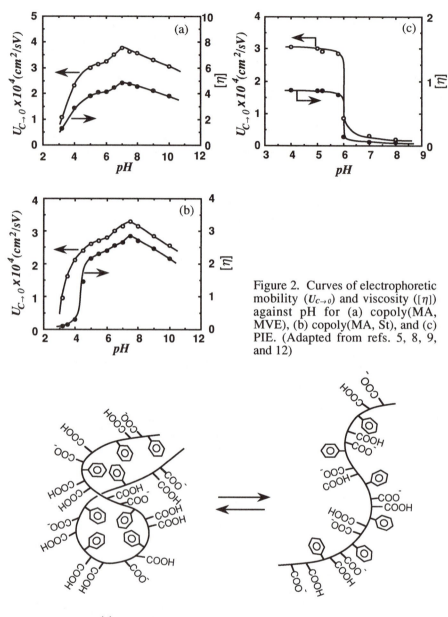

Figure 2. Curves of electrophoretic mobility ($U_{C \to 0}$) and viscosity ($[\eta]$) against pH for (a) copoly(MA, MVE), (b) copoly(MA, St), and (c) PIE. (Adapted from refs. 5, 8, 9, and 12)

Figure 3. Schematic representations for conformational transition of Copoly(MA, St): (a) a tightly coiled chain due to hydrophobic interaction between the phenyl groups; (b) an extended chain due to electrical repulsion between the charged carboxylic groups

extended one (pH < 6), which takes place via a quick change of polyion charges as characterized by the mobility curve. Hydrogen bondings between H_2N- and/or -NH- groups attached to the branched chain seem to act as an attractive force *(12)*.

In the case of PIE, however, both the mobility and viscosity curves increase rapidly at the same pH. This is different from the results for copoly(MA, St), which showed that the viscosity increases rapidly at pH 4.5 while the mobility increases slowly. Our previous studies *(8,13)* on electrophoresis and viscosity for different polyelectrolytes have shown that when ionic strength is greater than 0.1, the relationship between $[\eta]$ and $U_{\rightarrow 0}$ can be expressed as follows:

$$[\eta] = AU_{C \rightarrow O} M^a \tag{1}$$

Here, M represents the molecular weight of the polyelectrolyte, and a and A are empirical constants (A is independent of both M and ionic strength). When equation (1) is combined with the Flory-Fox equation *(14)* for the viscosity, $U_{\rightarrow 0}$ can be related to the expansion factor (α_η):

$$\left.\begin{array}{l} \alpha_\eta^3 = (A/K_O)(U_{C \rightarrow O})M^c \\[6pt] K_O = \phi\, (\overline{\gamma_0}^2/M)^{3/2} \\[6pt] c = a - 1/2 \end{array}\right\} \tag{2}$$

Here, ϕ is a universal constant relating to the end-to-end distance at the theta state. The plots of $[\eta]$ vs $U_{\rightarrow 0}$ for copoly(MA, MVE) and PIE from Figure 2 were expressed by straight lines passing through the origin (data not shown), as predicted by equation (1). This was also the case for copoly(MA, St) at pH > 5. Therefore, the pH-induced conformational changes of these polyelectrolytes can be understood as an alteration in the expansion factor due to the net charge density of the polyions, which can be shown by changes in $U_{\rightarrow 0}$. However, the plots of $[\eta]$ vs $U_{\rightarrow 0}$ for copoly(MA, St) at pH < 5 deviated from a straight line. It is thus likely that the "tightly coiled chains" of copoly(MA, St) and PIE are different from one another. This seems to be related to the following: (i) PIE is tightly coiled at pH > 6 due to hydrogen bondings between the amino and/or imino groups attached to branched chains, but (ii) copoly(MA, St) forms a coil at pH < 4 due to the hydrophobic interaction between the phenyl groups along the linear chains.

pH Dependence of Permeability of Microcapsule Membranes with a Polyion-Adsorbed Surface Layer

A polyion layer adsorbed onto a capsule membrane, as depicted in Figure 1, would result in a resistance to mass transfer. Thus, a study of the permeation of low molecular-weight solutes through such a capsule membrane would be useful for obtaining information on pH-induced changes in the configuration of the adsorbed polyion. In this section, the effects of pH on the permeability of microcapsules with adsorbed polyelectrolytes are investigated and compared with results from the viscometric and electrophoretic studies.

Materials and Methods. Stable microcapsules (mean diameter, 8 ~ 10 μm) with a semipermeable membrane of poly(styrene) (PSt) were prepared by depositing the polymer around emulsified aqueous droplets using the following three procedures: (i) primary emulsification of an aqueous solution of sodium dodecylbenzenesulfonate or Triton X-100 as an emulsifier in dichloromethane containing PSt with a homoblender;

(ii) secondary emulsification of the water/organic-type emulsion obtained in an aqueous solution containing either of the above emulsifiers under vigorous agitation; and (iii) complete removal of dichloromethane from the resulting (water/organic)/water-type complex emulsion according to the literature *(15)*. The microcapsules were collected by centrifugation, thoroughly washed with distilled water, and then subjected to the polymer adsorption procedure.

The polymer adsorption was performed by stirring the capsules in appropriate buffers containing polyelectrolytes at room temperature. Usually, the size of capsule suspensions was adjusted to 100 ~ 120 mL, and the stirring allowed to continue for 10 h. After adsorption, the capsules were recovered by centrifugation and purified by repeated washing with the buffer used in the adsorption until no polymer was detected in the washing extracts, as determined by colloid titration [ref. 16 for copoly(MA, MVE) and copoly(MA, St); ref. 17 for PIE].

Permeability was estimated by measuring the concentration changes of *n*-propyl alcohol (PA) as a permeate, after quick mixing of the proper quantity of aqueous PA solution and capsule suspension, both of which were previously kept at the same temperature (25 °C) and adjusted to the same pH (3 ~ 10) with 0.1M acetate, phosphate, or carbonate buffer. The samples (usually, 0.1 mL) were separated from the suspension using a 0.1 μm filter at suitable time intervals and analyzed with a Hitachi model 635A liquid chromatograph. The permeability constant, *P*, was then calculated using the following equation, derived from Fick's first law of diffusion *(15)*:

$$P = -\frac{C_f V_m}{C_i A t} \ln \frac{C_t - C_f}{C_i - C_f} \tag{3}$$

Here, C_i, C_t, and C_f are the initial, intermediary (at time *t*), and final concentrations of PA, respectively; V_m is the total volume of the microcapsule; and *A* is the total surface area of the microcapsules. The plot of $log[(C_t - C_f)/(C_i - C_f)]$ against time showed a straight line throughout all measurements. Typical examples of the linear plots obtained are represented in Figure 4.

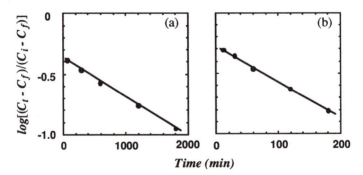

Figure 4. Plots of $log[(C_t - C_f)/(C_i - C_f)]$ against time for PSt microcapsules with adsorbed copoly(MA, St) at (a) pH 3.1 and (b) pH 7.1. (Adapted from ref. 5)

Results and Discussion. As shown in Table I, pH changes in *P* were affected by the amount of the polymer adsorbed onto the outer surface of the microcapsules. Thus, the amount of the adsorbed polymer was controlled by means of the initial concentrations of polyelectrolyte solution in the adsorption experiments; the optimal condition was found by trial and error, as is usual for any chemical reaction.

The curves of P vs pH for the three polyelectrolytes are shown in Figure 5, together with that for the original PSt capsules without any adsorbed polymer layer. In the case of the membrane of the original capsules, the P value was independent of pH and remained constant (1.59×10^{-5}). In contrast, the P values for the capsules with the adsorbed polyelectrolytes varied depending on pH. In the cases of the two MA copolymers, an increase in pH brought about an increase in permeability, whereas the permeability of the capsules with PIE increased with decreasing pH. Such permeability changes are analogous to the changes in the viscosity curves shown in Figure 2. In particular, the capsules with copoly(MA, St) and PIE exhibited a rapid change in permeability at the pH value at which the conformational transitions of both polyelectrolytes occurred in aqueous media. These results indicate that the configuration of the polyions adsorbed onto the PSt capsule membrane varies in similar ways in that they undergo conformational changes as a result of pH-induced electrostatic interactions.

Another important characteristic of the permeability changes observed here was that the value of P_{min} for the copoly(MA, St)-adsorbed capsules was much smaller than that for PIE-adsorbed capsules (see Table I), whereas the amount of the adsorbed PIE was larger than that of copoly(MA, St). As was discussed in the previous section, there is a difference in the tightness of the coiled chains of PIE and copoly(MA, St); that is, copoly(MA, St) seems to be more highly coiled than PIE. Thus, densely packed copoly(MA, St) coils could cover the surface of the capsules, leading to a reduced permeability despite the fact that the amount adsorbed is smaller than that of PIE. As a result, a study of the permeability of the capsule membrane with adsorbed polyelectrolytes permits an examination of the fine alterations in their configurations occurring on the membrane surface.

Application of Microcapsules with Adsorbed Polyelectrolytes in the On/Off Control of Enzyme Reactions

Capsule membranes with an adsorbed polyelectrolyte layer through which the permeation of solutes is dramatically altered in response to small changes in pH would be useful in constructing a functional encapsulated enzyme system in which the initiation/termination of an enzymatic reaction could be controlled. PSt microcapsules with copoly(MA, St) or PIE seem to be well suited for this purpose, since their permeability alters rapidly over a very narrow pH range as a result of changes in the configuration of the adsorbed polyion. The present section describes the on/off control of an enzyme reaction using pH-sensitive PSt microcapsules with copoly(MA, St).

Encapsulation of Enzyme. To prepare the enzyme-loaded PSt microcapsules, a primary emulsification was carried out on 20 mL of 0.1M acetate buffer containing 600 mg of enzyme (invertase) and 800 mg of an emulsifier (Triton X-100) *(5)*. The purification of the enzyme-loaded capsules, followed by polymer adsorption, was performed in the same manner as described in the previous section.

Initiation/Termination Control of Enzyme Reaction. The enzymatic hydrolysis of sucrose was studied at 25°C through a monitoring of the formation rate of reducing sugar products (an equimolar mixture of glucose and fructose) in an aqueous suspension of the enzyme-loaded capsules. The suspension (50 mL) initially contained 100 mM of the substrate and 13.5% (v/v) of the capsules with a total of 200 mg of the encapsulated enzyme. On/off control was tested by investigating the batch reaction kinetics at pH 5.5 (at which the reaction occurs) and at pH 4.5 (at which the reaction stops). The adjustment of pH was made by quick additions of a small amount of 2M HCl or NaOH.

Table I. Effect of amount of Polyelectrolytes adsorbed on Permeability

Polyelectrolyte	Adsorption Conditions		Adsorbed Amount ($\mu g/cm^2$)	Permeability [a,b] ($\times 10^{-6}$ cm/sec)	
	Initial Polym. Concn. (w/v %)	pH		P_{max}	P_{min}
PIE	0.1	4.0	4.5	16.2 (4)	6.6 (8)
PIE	1.0	4.0	24.3	14.0 (4)	2.7 (8)
PIE	1.0	8.0	23.3	14.6 (4)	2.0 (8)
PIE	10.0	4.0	235.8	1.1 (4)	0.6 (8)
Copoly(MA, MVE)	0.005	3.0	0.16	2.8 (8)	0.35 (3)
Copoly(MA, St)	0.05	3.0	1.7	2.55 (7)	0.062 (4)

[a] Value in parentheses represents the pH for the permeability measurement. [b] P_{max} and P_{min} denote maximum and minimum values obtained under the conditions of measurements, respectively.

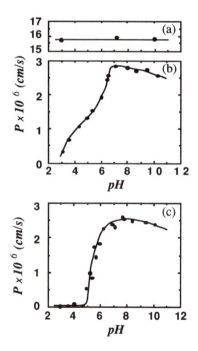

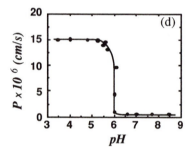

Figure 5. Curves of permeability constant (P) against pH for PSt microcapsules: (a) without adsorbed polyelectrolyte layer; with adsorbed layers of (b) copoly(MA, MVE), (c) copoly(MA, St), and (d) PIE. (Adapted from refs. 5 and 6)

Figure 6 shows a typical example of the on/off control of the enzymatic reaction. Using the original enzyme-loaded PSt microcapsules without the absorbed polymer layer, enzymatic hydrolysis occurred not only at pH 5.5 but also at pH 4.5, resulting in the formation of both glucose and fructose. In contrast, when the capsules with the adsorbed copoly(MA, St) were employed at pH 4.5, the catalytic action of the loaded enzyme was nearly or entirely depressed (the concentration of the reducing sugars produced was less than 0.1 μg/mL), but the reaction could be initiated by adjusting the pH of the outer medium to 5.5. Such on/off control could be repeated reversibly throughout a single run of measurements. In addition, repeated measurements over several days gave excellent reproducibility without damage to the capsules.

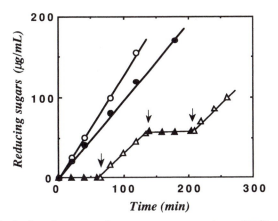

Figure 6. Hydrolysis of sucrose in aqueous suspension of PSt microcapsules without (●, ○) and with (▲ ,△) the adsorbed layer of copoly(MA, St) at (●,▲) pH 4.5 and (○,△) pH 5.5. The arrows show pH adjustment by quick addition of a small amount of 2M HCl or NaOH to the capsule suspension. (Adapted from ref. 5)

Similar pH-sensitive on/off control of the hydrolysis of maltotriose can be performed using β-amylase-loaded PSt microcapsules with a surface layer of adsorbed PIE (6). Therefore, pH-sensitive microcapsules, the outer surfaces of which are covered with the adsorbed polyions undergoing rapid changes in configuration, enable the control of enzymatic processes through small pH changes in the outer medium.

Conclusions and Future Topics

The main conclusions to be drawn from the present study are as follows: (i) The examination of the permeability of microcapsules with a layer of adsorbed polyelectrolyte is a useful method for obtaining information on the configurational changes of polyions on polymer membranes. (ii) A dramatic change in the configuration of copoly(MA, St) or PIE adsorbed onto the surface of a PSt capsule membrane occurs over a pH range in which these polyions undergo a conformational transition in aqueous solution. (iii) The configuration of copoly(MA, MVE) on a capsule membrane varies gradually depending on the pH of the outer media, which is also comparable with the conformational change in an aqueous solution. (iv)

Microcapsules with a surface layer of copoly(MA, St) or PIE permit the control of enzymatic process through small pH changes in the outer medium.

Certain other polymers are currently available for the preparation of microcapsules *(18-20)*. Configurational changes of the polyelectrolytes adsorbed, not only onto a hydrophobic PSt membrane but also onto polymer or copolymer membranes with different hydrophobicities, can be thus studied using the present method. Another interesting topic for future study is the use of polypeptides and proteins instead of synthetic polyelectrolytes. Such systematic studies would provide a good basis for understanding the conformational and/or configurational changes of protein molecules on biological membranes in living systems.

In addition to these basic subjects, a microcapsule with a layer of polyelectrolyte, the configuration of which is dramatically altered in response to small changes in the pH of the outer medium, permits the on/off control of enzymatic processes. Such microencapsulated enzyme systems, with expected applications in biochemical sensors and display devices, constitute "functional" immobilized enzymes that surpass the conventional definition, or usually credited advantages, of immobilized enzymes (Kokufuta, E. *Prog. Polym. Sci.*, in press). Thus, the encapsulation of enzymes using pH-sensitive microcapsules with adsorbed polyelectrolytes, the permeability of which can be controlled through small differences in pH over a wide range of pH values, would be of technical interest in the developing interdisciplinary areas of enzymology, polymer chemistry, and biomedical engineering.

Acknowledgments

The author wishes to thank Dr. T. Sodeyama for valuable discussions and R. Craig for his critical reading of this manuscript. In addition, the author is grateful for T. Katano and N. Shimizu for their assistance in the primary experiment of the series included in the present study. This work was supported by Grants-in-Aid for Scientific Research from the Ministry of Education and New Energy Development Organization (NEDO), Japan.

Literature Cited

1. Kokufuta, E.; Takahashi, K. *Macromolecules* **1986**, *19*, 351.
2. Kokufuta, E.; Fujii, S.; Hirai, Y.; Nakamura, I. *Polymer* **1982**, *23*, 452.
3. Kokufuta, E.; Fujii; S.; Nakamura, I. *Makromol. Chem.* **1982**, *183*, 1233.
4. Kokufuta, E.; Hirai; Y.; Nakamura, I. *Makromol. Chem.* **1981**, *182*; 1715.
5. Kokufuta, E.; Shimizu, N.; Nakamura, I. *Biotechnol. Bioeng.* **1988**, *32*, 289.
6. Kokufuta, E.; Sodeyama, T.; Katano, T. *J. Chem. Soc., Chem. Commun.* **1986**, 641.
7. Ilmain, F.; Tanaka, T.; Kokufuta, E. *Nature* **1991**, *349*, 400.
8. Kokufuta, E. *Polymer* **1980**, *21*, 177.
9. Kokufuta, E.; Ito, M.; Hirata, M.; Iwai, S. *Kobunshi Ronbunshu (Jpn. Edn.)* **1974**, *31*, 688; *Kobunshi Ronbunshu (Engl. Edn.)* **1974**, *3*, 1957.
10. Ohno, N.; Nitta, K.; Makino, S.; Sugai, S. *J. Polym. Sci. (Polym. Phys. Edn.)* **1973**, *11*, 413.
11. Ferry, J. D.; Udy. D. C.; Wu, F. C.; Heckler, G. E.; Fordynce, D. B. *J. Colloid Sci.* **1951**, *6*, 429.
12. Kokufuta, E.; Hirata, M.; Iwai, S. *Kobunshi Ronbunshu (Jpn. Edn.)* **1974**, *31*, 234; *Kobunshi Ronbunshu (Engl. Edn.)* **1974**, *3*, 11383.
13. Kokufuta, E.; Hirata, M.; Iwai, S. *Nippon Kagaku Kaishi* **1975**, 369 (in Japanese).

14. Flory, P. J.; Fox, T. G. *J. Amer. Chem. Soc.* **1951**, *73*, 1904.
15. Kitajima, M.; Kondo, A. *Bull. Chem. Soc. Jpn.* **1971**, *44*, 3201.
16. Kokufuta, E.; Iwai, S. *Bull. Chem. Soc. Jpn.* **1977**, *50*, 3043.
17. Kokufuta, E. *Macromolecules* **1979**, *12*, 350.
18. Kokufuta, E. *Fragrance J.* **1991**, *19*, 16 (in Japanese).
19. *Microcapsules and Other Capsules*; Gutcho, M. H., Ed.; Noyes Data Co.: Park Ridge, NJ, 1979.
20. Kondo, A.; *Microcapsule Process and Technology*, Marcel Dekker: New York, 1979.

RECEIVED February 23, 1993

Chapter 9

Light-Scattering Studies of Adsorption of End-Functionalized Polymers in Colloidal Solutions

Penger Tong[1], B. L. Carvalho[2], J. S. Huang[3], and L. J. Fetters[3]

[1]Department of Physics, Oklahoma State University, Stillwater, OK 74078
[2]Department of Materials Science and Engineering, Massachusetts Institute of Technology, Cambridge, MA 02139
[3]Exxon Research and Engineering Company, Annandale, NJ 08801

We report a light scattering study of the adsorption of end functionalized polymers on colloidal spheres. A scattering method is developed to measure the fraction of polymer molecules adsorbed on the colloidal surfaces. Measurements of the scattering intensity as a function of the polymer-to-colloid molar ratio reveal that only a fraction of the end-functionalized polymers adsorbs on the colloidal surface. The results for the end-functionalized polymers are compared with those for the un-functionalized polymer. It is found that the interaction between the colloid and the un-functionalized polymer is repulsive, which introduces a depletion-attraction between the colloidal particles. The functional end-groups are found to interact attractively with the polar cores of the colloidal particles. The adsorption energy between the functional group and the colloidal surface is estimated to be $\sim 4k_BT$. The presence of the adsorbed polymer on the colloidal surfaces greatly reduces the depletion attraction, and therefore, enhances the stability of the colloid-polymer mixture.

The novel interactions between polymers and colloidal surfaces have spawned a variety of applications. Paint, for example, is often stabilized through the adsorption of some polymer onto the colloidal pigment. Water, on the other hand, can be rid of colloidal impurities by adding a small amount of polymer, allowing the colloidal particles to aggregate and then filtering the solution. These seemingly disparate technologies are both the consequence of the entropic requirements for the polymer molecules in colloidal solutions. In the case of colloidal stabilization, the adsorbed polymer resists the approach of other surfaces through a loss of its conformational entropy. Surfaces, then, are maintained at separations large enough to damp any attractions and the colloidal suspension is stabilized. The adsorbed polymer layer affects not only the thermodynamics but also the hydrodynamics of the colloidal suspension.[1,2] Dynamic light scattering has been

0097–6156/93/0532–0096$06.00/0

used to estimate the apparent hydrodynamic thickness of an adsorbed polymer layer.[3] The hydrodynamic thickness is the difference between the Stokes' radius of the bare colloidal particle in the solvent alone and the value for the particle in the polymer solution. While it is straightforward to perform a dynamic light scattering measurement on a mixture of colloid and polymer, any determination of the hydrodynamic thickness requires all the polymer molecules to stick on the colloidal surfaces.

Consider now the polymer-induced attraction of bare colloids. When two colloidal particles approach each other, free polymer chains in the inter-particle gap resist a reduction in their conformational entropy and escape from the neighborhood of the colloidal particles. The colloidal particle is then surrounded by a depletion zone with a polymer concentration substantially lower than the bulk concentration of polymer. In an effort to reduce the free energy of the system, the colloidal particles come together and share their depletion volumes. This results in an effective attraction between the colloidal particles.[4] If the attraction is large enough, phase separation or flocculation of the colloidal particles occurs. The depletion effect was first recognized by Asakura and Oosawa,[5] and in recent years, many theoretical and experimental studies of the depletion effect have been carried out in various aqueous and organic colloidal solutions. Most experimental studies, however, are restricted to examining the phase behaviour of the colloid-polymer mixtures. Recently, we developed a light scattering approach to probe changes of the interaction potential between the colloidal particles in a free polymer solution.[6] In our experiment, the second virial coefficient of the colloidal particles as a function of the free-polymer concentration was obtained from measurements of the concentration dependence of the light intensity scattered from the mixture. The experiment demonstrates that the light scattering scheme is indeed capable of measuring the depletion effect in the colloid-polymer mixture.

In this paper we report a static light scattering study of the adsorption of end-functionalized polymers on colloidal spheres. Monodispersed hydrogenated polyisoprene and its single-end-functionalized derivatives are used to modify the interaction between the polymer and the colloid. The experiment reveals that only a fraction of the end-functionalized polymers adsorbs on the colloidal surface. The adsorption energy between the functional group and the colloidal surface is estimated to be $\sim 4k_BT$. The results for the end-functionalized polymers are compared with those for the un-functionalized polymer. Because the un-functionalized polymer does not adsorb, at all, to the colloidal spheres, we view the adsorption of the end-functionalized polymer as occurring through the end-group. It is found that the polymer adsorption onto the colloidal surfaces greatly reduces the depletion attraction between the colloidal particles. The experiment is of interest to observe the microscopic interaction between colloid and polymer and to see how it responds to the incorporation of functional group on the polymers. With this knowledge, one can estimate the phase stability properties of colloid-polymer mixtures in a straightforward way.

Theory

Scattering from a Mixture of Colloid and Polymer. The scattering intensity from a mixture of colloid and polymer can be written as[6]

$$BM_1\rho_1'/R(0) = Y(\rho_2') + \frac{2\rho_1'}{M_1}P(\rho_2'), \tag{1}$$

where the intercept

$$Y(\rho_2') = 1 - \frac{\rho_2'}{M_2}\frac{2f_2}{f_1}C_{12} - (\frac{\rho_2'}{M_2})^2\{\frac{2f_2}{f_1}(C_{12}C_{22} + E_2)$$

$$-(\frac{f_2}{f_1})^2(3C_{12}^2 - B_1)\} + \mathcal{O}(\rho_2^3), \tag{2}$$

and the slope

$$P(\rho_2') = -\frac{C_{11}}{2} - \frac{\rho_2'}{2M_2}\{(C_{12}^2 + A_2) - \frac{2f_2}{f_1}(C_{12}C_{11} - E_1)\} + \mathcal{O}(\rho_2^2, \rho_1). \tag{3}$$

In the above, M, ρ' and f are the molecular weight, the mass density (gm/cm^3) and the scattering amplitude, respectively. Here the colloidal particle is denoted as component 1 while the polymer molecule is denoted as component 2. The constant $B = B'(f_1/M_1)^2$, with B' being an instrumental constant. The excess intensity $R(0)$ is defined as $I(0) - I_0$, where I(0) is the scattered intensity from the mixture at concentrations ρ_1' and ρ_2', measured at the scattering angle $\theta = 0$, and I_0 is the scattering intensity of the polymer solution alone $(\rho_1' = 0)$. The coefficients in Eqs. (2) and (3) are the density-expansion coefficients for the partial structure factor[7] S_{ij}, with $-C_{ij}/2$ being the second virial coefficients and A_2, B_1, E_1, and E_2 being the higher-order expansion coefficients. These coefficients have been calculated in Ref. 6.

If the polymer is invisible $(f_2 = 0)$, $P(\rho_2')$ in Eq. (3) becomes an effective second virial coefficient $b_{11}(\rho_2')$ of the colloidal particles at a given polymer concentration ρ_2'. This virial coefficient has the usual interpretation in terms of osmotic pressure derivatives. Equation (3) states that the interaction between a colloidal particle and a polymer molecule reduces the value of $b_{11}(\rho_2')$, and therefore the effective interaction between the colloidal particles may become attractive if enough polymer is added. When the polymer is visible $(f_2 \neq 0)$, the interference between the two species changes both the intercept Y and the slope P. From the measured $Y(\rho_2')$ the colloid-polymer interaction parameter C_{12} can be obtained using Eq. (2). Therefore, we can find quantitatively how the two species attract or repel. Experimentally, a straight line can be obtained at low ρ_1' end when the measured $\rho_1'/R(0)$ is plotted against ρ_1'. From the intercept and the slope of the straight line one obtains C_{12} and $b_{11}(\rho_2')$.

The two independent variables used in Eq. (1), which determine the total scattering intensity of the colloid-polymer mixture, are the concentrations of colloid ρ_1' and polymer ρ_2'. Sometimes one may find it more convenient to work with the polymer-to-colloid molar ratio $\omega = \rho_2/\rho_1$, instead of ρ_2. When ρ_2 is replaced by $\omega\rho_1$, Eqs. (2) and (3) become

$$Y(\omega) = \frac{1}{1 + (f_2/f_1)^2\omega}, \tag{4}$$

and

$$P(\omega) = -\frac{C_{11} + 2C_{12}(f_2/f_1)\omega + C_{22}(f_2/f_1)^2\omega^2}{2(1 + (f_2/f_1)^2\omega)^2} + \mathcal{O}(\rho_1). \tag{5}$$

In the above discussion we have assumed that the polymer molecules are free in the solution, so that they scatter light individually. When the end-functionalized polymer is added into the colloidal suspension, some of the polymer molecules adsorb on the surface of the colloidal particle, and form colloid-polymer micelles. If all the polymer molecules adsorb on the colloidal surfaces (complete adsorption), the mixture can be viewed as a single-component system consisting of only colloid-polymer micelles. In this case the standard virial expansion for the scattering light intensity is still valid, and Eq. (1) becomes[6]

$$BM_1\rho_1'/R(0) = \frac{1}{(1 + (f_2/f_1)\omega)^2}\{1 - \frac{\rho_1'}{M_1}C_{33}(0) + \mathcal{O}(\rho_1^2)\}. \tag{6}$$

In the above, $-C_{33}(0)/2$ is the second virial coefficient for the colloid-polymer micelles, and the scattering amplitude of the colloid-polymer micelle is assumed to be $f_1 + \omega f_2$, with f_1 and f_2 being the scattering amplitudes for the colloidal particle and the polymer molecule, respectively. One can immediately see that the intercept $Y(\omega)$ in Eq. (6) decays much faster than that in Eq. (4) ($f_2/f_1 \simeq 0.1$ for our colloid-polymer mixture). This difference in $Y(\omega)$ reflects the fact that the scattering from the colloid-polymer micelles is coherent, and it is incoherent if colloidal particles and polymer molecules are independent with each other. The slope $P(\omega)$ is a measure of the "interaction volume" of the mixture. When there is no adsorption in the colloid-polymer mixture, the total "interaction volume" of the mixture is a intensity normalized sum of the individual excluded volumes (i.e. the second virial coefficients) of a colloidal particle and a polymer molecule as well as the excluded volume between the two species. For the complete adsorption, on the other hand, the "interaction volume" is just the excluded volume of the colloid-polymer micelle.

When the adsorption energy ϵ between the end-functional group and the colloidal surface is not much larger than k_BT, the polymer molecules partition themselves between the bulk fluid and the adsorbed state. If one denotes the number fraction of the adsorbed polymer by α, the number fraction of the free polymer is then $1 - \alpha$. It will be shown in the next section that the partition coefficient α is proportional to the amount of colloid and depends exponentially on the free energy of adsorption. The partition between free and the adsorbed polymers is very important in determining the adsorption energy ϵ (see next section). In the partial adsorption case, the colloid-polymer mixture can be viewed as a two-solute system consisting of the colloid-polymer micelles and free polymer molecules. Equation (1) can still be used to calculate the total scattering intensity of the mixture, except one has to use $f_3 = f_1 + \alpha \omega f_2$ as the scattering amplitude of the colloid-polymer micelle, and $(1 - \alpha)\rho'_2$ as the free polymer concentration. At the infinite dilution of the colloidal concentration, the intercept $Y(\omega)$ has the same functional form as that in Eq. (4). This is because α approaches zero as the colloidal concentration goes to zero.

At a finite colloid concentration $\rho'_1 = \rho'_0$, the scattering intensity in the colloid-polymer mixture can be written as

$$BM_1 \rho'_0 / R(0) = Y(\omega, \alpha_0) + (2\rho'_0/M_1)P(\omega, \alpha_0), \tag{7}$$

where the "intercept"

$$Y(\omega, \alpha_0) = \frac{1}{(1 + \alpha_0 \omega \frac{f_2}{f_1})^2 + (1 - \alpha_0)\omega(\frac{f_2}{f_1})^2}, \tag{8}$$

and the slope $P(\omega, \alpha_0)$ has the same form as that in Eq. (5) when ρ'_0 is small. In the above, α_0 is the polymer partition coefficient at ρ'_0. The first term in Eq. (7) is the contribution from the increment of the scattering amplitude of the colloid-polymer micelle as ρ'_2 is increased. The second term is due to the interactions in the mixture. As will be shown in the experiment, at low colloid concentrations the change of the scattering light intensity is predominantly from the "intercept" $Y(\omega, \alpha_0)$, from which the polymer partition coefficient α_0 at ρ'_0 can be obtained. The "intercept" $Y(\omega, \alpha_0)$ as a function of ω lies in-between the two curves of $Y(\omega)$ for the non-adsorption case (Eq. (4)) and for the complete adsorption case (Eq. (6)), as one might expect.

Estimation of the Adsorption Energy ϵ. The partition coefficient α discussed above is the probability for a single polymer chain to anchor to the colloidal surface. In doing so the system reduces its free energy. Therefore, the fraction of the adsorbed polymer is

$$\alpha = \frac{1}{1 + e^{-\Delta F/k_B T}}, \tag{9}$$

where k_B is Boltzman's constant, T is the absolute temperature, and ΔF is the free energy gain for a free polymer chain to adsorb on the colloidal surface. In the experiment to be discussed below, both the free polymer concentration and the surface coverage of the adsorbed polymer are below the overlap polymer concentration, so that one can ignore the excluded volume interactions between the polymer chains. In this dilute limit, the free energy gain can be written as

$$\Delta F = \epsilon - T\Delta S. \tag{10}$$

In the above, ϵ is the adsorption energy between the end-functional group and the colloidal surface, and ΔS is the entropy loss for a free polymer chain to anchor to the colloidal surface.

When a polymer chain adsorbs on a colloidal surface, it loses its entropy in two ways. First, there is a loss of conformational entropy that the polymer chain experiences when one of its ends is fixed. The total number of configurations for a free polymer chain in a good solvent scales[8] as $\tilde{z}^N N^{\gamma-1}$, where N is the degree of polymerization, $\tilde{z}$ is an effective coordination number, and $\gamma = 7/6$ for three-dimensional lattices. A computer simulation by Eisenriegler et al.[9] has shown that for a polymer chain with its one-end fixed at a flat wall, otherwise in a good solvent, the number of configurations has the same scaling form as for a free polymer chain but with $\gamma = 0.695$. Therefore, the probability of finding a polymer chain with its one-end fixed at a flat wall is $\Psi(N) = \Psi_0 N^{-\gamma'}$, where the exponent $\gamma' = 7/6 - 0.695 = 0.472$. The proportionality constant Ψ_0 should be of order unity because $\Psi(N = 1) = 1$. The change of conformational entropy then has the form

$$\Delta S_{conf} = -k_B ln \Psi(N) = 0.472 k_B ln(N). \tag{11}$$

The adsorption of polymer molecules onto the colloid also costs translational entropy. The loss of translational entropy can be expressed as

$$\Delta S_{trans} = -k_B [ln(\frac{3\phi_1 \delta}{R_{11}})], \tag{12}$$

where R_{11} is the radius of the colloidal particles, ϕ_1 is their volume fraction, and δ is the layer thickness of the adsorbed polymer. Incorporating the two entropic losses into Eq. (9), one can solve for the adsorption energy ϵ, that must be supplied to the polymer in order to adsorb on the colloidal surfaces,

$$\epsilon = k_B T \{0.472 ln(N) - ln(\frac{3\phi_1 \delta}{R_{11}}) - ln(\frac{1-\alpha}{\alpha})\}. \tag{13}$$

In the case where the un-adsorbed polymer molecules in the solution form polymer aggregates (see the discussion for the zwitterion-PEP below), Eq. (13) has the form

$$\epsilon - f_0 = k_B T \{0.472 ln(N) - ln(\frac{3\phi_1 \delta}{R_{11}}) - ln(\frac{1-\alpha}{\alpha})\}. \tag{14}$$

In the above, $f_0(p, N)$ is the free energy of a single polymer chain in a polymer aggregate with p being the aggregation number and N being the degree of polymerization of the polymer chains. Bug et. al.[10,11] have calculated $f_0(p, N)$ for spherical polymer aggregates. Here we treat f_0 as a constant because there are some unknown numerical parameters[10] in estimating f_0.

Experimental

The colloidal particle chosen for the study consists of a calcium carbonate core ($CaCO_3$) with an adsorbed monolayer of a randomly branched calcium alkyl-benzene sulphonate (CaSA) surfactant. The monolayer has a thickness of[12]

$19 \pm 1 \overset{\circ}{A}$. The colloidal particles are dispersed in decane. The synthesis procedures used to prepare the colloidal dispersion have been described by Markovic et al.[12] These colloidal particles have been well characterized previously using small-angle neutron and light scattering techniques,[6,12] and are used as an acid-neutralizing aid in lubricating oils. Such a non-aqueous dispersion is ideal for the investigation attempted here since the colloidal system is approximately a hard-sphere system.[13] Our dynamic light scattering measurements[6] revealed that the colloidal particles had a hydrodynamic radius of 5.0 nm and that the size polydispersity was approximately 10 %. The molecular weight of the colloidal particle $M_1 = 300,000 \pm 15\%$, which was obtained from a sedimentation measurement.[6]

The polymer used in the study was hydrogenated polyisoprene, i.e. alternating poly(ethylene-propylene) (PEP) and its single-end-functionalized derivatives, which were synthesized by the anionic polymerization scheme.[14,15] One derivative contains a tertiary amine group capped at one end of the chain (amine-PEP). The second has a strongly polar sulphonate-amine zwitterion at the end of the chain (zwitterion-PEP). The parent PEP and its end-functionalized derivatives are model polymers, which have been well characterized previously using various experimental techniques.[14,15] The ratio M_w/M_n was well below 1.1 for samples in the study.

Results and Discussion

Figure 1a presents the scattering data for the PEP with the molecular weight $(M_2)_f = 26,000$ (solid circles) and for the amine-PEP with $(M_2)_a = 25,000$ (open circles). (A subscript outside the parentheses is used to identify a quantity, which is related to different polymers. Letters f, a and z are used for <<free>>, <<amine>> and <<zwitterion>>, same afterwards). The data is plotted as $K_2 \rho_2' / (I(\theta)/I_0 - 1)$ versus ρ_2', where $I(\theta)$ is the scattering intensity from the polymer solution at the concentration ρ_2' (gm/cm^3), measured at the scattering angle θ, and I_0 is the light intensity scattered from the solvent alone $(\rho_2' = 0)$. The constant K_2 in the plot was chosen such that $K_2 \rho_2' / (I(\theta)/I_0 - 1) = 0$ when $\rho_2' = 0$. Since the size of the polymer molecules (see Table I) is much smaller than the wavelength of the incident light $(\lambda = 623.8 \; nm)$, the measured $I(\theta)$ is independent of θ. Figure 1a shows that the two sets of data are almost the same, indicating that there is no association for the amine-PEP polymer. This conclusion was also drawn by Davidson et. al. in a study of the association behavior of the same end-functionalized polymers.[15] The value of b_{22}/M_2 (b_{22} and M_2 being the second virial coefficient and the molecular weight of the polymer molecules, respectively) is determined from the slope of the fitted straight line. With the measured b_{22} one can define the radius R_{22} of an equivalent hard sphere such that $4(4\pi/3)R_{22}^3 = b_{22}$. In general, $R_{ij} = 0.39(b_{ij})^{1/3}$, where the subscript ij denotes the species. The zwitterion-PEP polymer chains are found to associate strongly in decane. Figure 1b shows the scattering data for zwitterion-PEP with $(M_2)_z = 25,000$. The constant $(K_2)_z$ in the plot is found to be much larger than $(K_2)_a$. From the ratio of the two constants (see Table I), we find the average

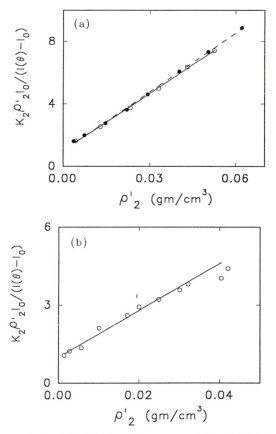

Figure 1. Plots of $K_2\rho_2'/(I(\theta)/I_0 - 1)$ versus ρ_2' for PEP (closed circles in Fig. 1a), amine-PEP (open circles in Fig. 1a), and zwitterion-PEP (open circles in Fig. 1b) in decane. The solid and dashed lines are linear fits to the data points. The scattering angle $\theta = 90^0$.

association number $n = (K_2)_z/(K_2)_a = 11.3$. Similar association behaviors for the zwitterion-PEP were also found in other organic solvents.[15] Table I lists the measured values of b_{ii}/M_i and the corresponding hard-sphere radii for our polymers and the colloid in decane.

Table I. Characterization of the colloidal particle and the polymers in decane ($i = 1$ for colloid and $i = 2$ for polymer)

Samples	b_{ii}/M_i (cm^3/gm)	R_{ii} (nm)	$K_i(\times 10^2)$	f_2/f_1
$M_1 = 300,000$	3.8	4.8	8.1	
$(M_2)_f = 26,000$	62.8	5.4	1.32	0.11
$(M_2)_a = 25,000$	61.2	5.3	1.54	0.13
$(M_2)_z = 25,000$	512.0	10.8	17.45	1.43

We now discuss the mixtures of the colloid and the polymers. Figure 2 shows plots of $K_1\rho'_1/(I(\theta)/I_0 - 1)$ versus ρ'_1 for the colloidal mixtures with different polymers in decane. The constant K_1 in the plot was chosen such that $K_1\rho'_1/(I(\theta)/I_0 - 1) = 0$ when $\omega = 0$. Figure 2a shows the effect of adding PEP at different molar ratios: $\omega = 3$ (open circles), $\omega = 4.82$ (closed circles), and $\omega = 15.3$ (open triangles). When the PEP is absent ($\omega = 0$), the plot is a straight line (see Fig. 2b). From the slope of the straight line one obtains the second virial coefficient b_{11} for the colloidal particles. When PEP is added to the colloidal suspension, the colloidal particles experience an attraction due to the depletion effect, as discussed in the theory section. This attraction shrinks the linear region in the virial expansion. Therefore, the plot of $K_1\rho'_1/(I(\theta)/I_0 - 1)$ versus ρ'_1 becomes curved. Figure 2a shows that the linear region in the plot becomes smaller with an increasing ω, which indicates that the polymer-induced attraction between the colloidal particles is increased. The initial slope of the plot is also increased with ω, a characteristic that is predicted by Eq. (5).

The curvature effect does not appear when the amine-PEP is added to the colloidal solution. This indicates a suppression of the depletion-induced attraction between the colloidal spheres. In fact, the colloidal suspension is stabilized by the adsorption of the amine-PEP polymer onto the colloidal surfaces, as we will discuss below. Figure 2b shows the scattering data measured in the colloid/amine-PEP mixture at three molar ratios: $\omega = 0$ (open circles), $\omega = 2.15$ (closed circles), and $\omega = 9.45$ (triangles). The solid lines are the linear fits to the data points: $1 + 10.0\rho'_1$ (top), $0.87 + 8.2\rho'_1$ (middle), and $0.57 + 7.4\rho'_1$ (bottom). Another striking feature of Figure 2 is that the scattering intensity at the smallest colloidal concentration $\rho'_0 = 0.01$ gm/cm^3 varies considerably when the

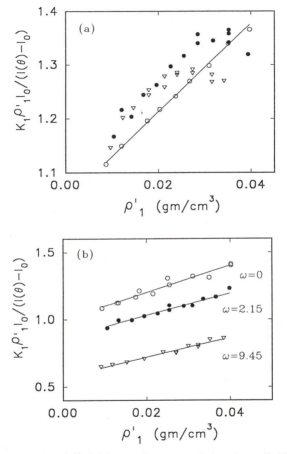

Figure 2. Plots of $K_1\rho_1'/(I(\theta)/I_0 - 1)$ versus ρ_1' for the colloidal mixtures with different polymers in decane at $\theta = 90^0$. (a) The colloid/PEP mixture with $\omega = 3$ (open circles), $\omega = 4.82$ (closed circles), and $\omega = 15.3$ (open triangles). The solid line is a linear fit to the data points with $\omega = 3$. (b) The colloid/amine-PEP mixture with $\omega = 0$ (open circles), $\omega = 2.15$ (closed circles), and $\omega = 9.45$ (triangles). The solid lines are the linear fits to the data points.

amine-PEP is added to the colloidal suspension (see Fig. 2b) but hardly at all for the colloidal mixture with PEP (see Fig. 2a).

In Fig. 2 the measured $BM_1\rho_1'/R(0)$ is plotted as a function of ρ_1', when ρ_1' is in the range between 0.01 gm/cm^3 and 0.04 gm/cm^3. The value of $BM_1\rho_1'/R(0)$ at ρ_0' ($= 0.01$ gm/cm^3) has been calculated in Eq. (7). From the linear fits in Fig. 2b we find that $(2\rho_0'/M_1)P(\omega,\alpha_0)$ is very small compared with $Y(\omega,\alpha_0)$ ($\sim 10\%$ of $Y(\omega,\alpha_0)$). Therefore, one can estimate α_0 from the measured "intercept" $Y(\omega,\alpha_0)$ using Eq. (8), and ignore the contribution from $(2\rho_0'/M_1)P(\omega,\alpha_0)$. Figure 3 shows the measured $Y(\omega)$ as a function of ω for the colloid/PEP mixture (close circles in Fig. 3a) and the measured $Y(\omega,\alpha_0)$ at a fixed $\rho_0' = 0.01$ gm/cm^3 for the colloidal mixtures with amine-PEP (open circles in Fig. 3a) and with zwitterion-PEP (open circles in Fig. 3b). For the un-functionalized PEP the measured $Y(\omega)$ is almost a constant, whereas for the two end-functionalized polymers $Y(\omega,\alpha_0)$ decreases with increasing ω. The large decrease in $Y(\omega,\alpha_0)$ indicates an adsorption of the polymer molecules onto the colloidal surfaces. This is because the colloid-polymer micelle scatters much more light than the colloid and polymer in their un-associated state.

The upper solid curve in Fig. 3a is a plot of Eq. (4) with $f_2/f_1 = 0.11$. The equation for the non-adsorption mixture fits the data well. Our previous study of the same system also suggests that the PEP polymer does not adsorb onto the colloidal spheres.[6] The two end-functionalized polymers are found to be partially adsorbed onto the colloidal surfaces. The lower solid curve in Fig. 3a is a fit to Eq. (8) with $\alpha_0 = 0.21$. In the range $0 < \omega < 24$, α_0 is found to be independent of the molar ratio ω. In this range of ω, the overall polymer concentration ρ_2' is below the polymer overlap concentration. The zwitterion-PEP data can also be fitted to Eq. (8) with a constant $\alpha_0 = 0.18$. The dashed curve in Fig. 3b is a plot of Eq. (8) with $\alpha = 1$ (complete adsorption). One can immediately see from Fig. 3 that the measured $Y(\omega,\alpha_0)$ for our end-functionalized polymers lies in-between the non-adsorption and complete adsorption curves.

With the fitted values of α_0, one can obtain the adsorption isotherm, which is a plot of the number of the adsorbed polymer molecules per colloidal particle ($\alpha_0\omega$) versus the polymer-colloid molar ratio ω. Because α_0 is a constant independent of ω in our working range of ω, the adsorption isotherm is then a simple linear function of ω, with the slope being α_0. The maximum value of $\alpha_0\omega$ obtained for our end-functionalized polymers is 5.4, which is smaller than the geometrical packing limit for the polymer molecules whose size is comparable to that of the colloidal particles (the coordination number is 8 for a simple body centered cubic lattice). In this weak adsorption limit, the polymer molecules partition themselves between the bulk fluid and the adsorbed state. The partition coefficient α is proportional to the amount of colloid and depends exponentially on the free energy of adsorption (see Eq. (13)). The configuration of the adsorbed polymer on the colloidal surface is expected to be approximately a random coil with one end stuck on the surface. In the opposite limit where the amount of polymer is much larger than that of colloid, we expect the adsorbed amount of polymer per colloidal particle to reach some saturating value.

At some point further attachment of polymer chains to colloid is inhibited by a lack of attachment sites or by repulsion from the already attached polymer molecules.

With the obtained values of α_0, the adsorption energy, ϵ, can be calculated using Eqs. (13) and (14). It is found that $\epsilon \simeq 4.3k_BT$ for the amine-PEP. In the calculation of ϵ, we have taken $R_{11} = 4.8$ nm, $\delta = 2R_{22} = 10.6$ nm, $\phi_1 = 8.3 \times 10^{-3}$, and $N = 25000/70 = 357$, where the monomer weight of amine-PEP is 70. For the zwitterion-PEP the obtained free energy difference $\epsilon - f_0 \simeq 4.1k_BT$, which is approximately the same as that for the amine-PEP. In the previous study[6] of interactions in the mixture of the colloid and the un-functionalized PEP, we found that the interaction between the colloid and the PEP is repulsive. The present study considers the effect of polar end groups on the polymer chain. In hydrocarbon solvents such end groups are expected to reduce the solubility of the polymers. In fact, we find that the zwitterion-PEP forms polymeric micelles in decane. In addition, the polar groups are found to interact attractively with the polar cores of the colloidal particles. The attractive potential energy ϵ is approximately $4k_BT$. The obtained adsorption energy, $\epsilon - f_0$, between the zwitterion group and the colloidal surface is small compared with that when the colloidal surface is replaced by a bare mica surface.[16] In the latter case the adsorption energy difference $\epsilon - f_0$ was estimated[17] to be $9k_BT$. The weaker adsorption is expected because the colloidal surface is coated with a monolayer of surfactants.

Our light scattering scheme is not only capable of measuring the amount of the polymer adsorbed on the colloidal surface, but also can probe changes of the microscopic interaction between the colloidal particles due to the polymer adsorption. Figure 4 shows the measured $P(\rho_2')/M_1$ as a function of the polymer concentration ρ_2' for PEP (closed circles) and amine-PEP (open circles). As mentioned in the theory section (see the discussion about Eq. (3)), $P(\rho_2')$ is related to the effective second virial coefficient for the colloidal particles in the polymer solution. The lower solid line in Fig. 4 is a fit to the linear function $3.4(1 - 51\rho_2')$. The intercept of the linear function is just the second virial coefficient $b_{11}(0)/M_1$ measured at $\rho_2' = 0$ (see Eq. (3)). We have calculated $P(\rho_2')$ using a binary hard-sphere model and obtained[6] $P(\rho_2')/b_{11}(0) = (1-46\rho_2')$, which agrees well with the measurement. The data in Fig. 4 thus demonstrate that the binary hard-sphere model[4,6,18] can indeed describe the depletion effect in our colloid/PEP mixture.

For the colloid/amine-PEP mixture, one can view it as a two-solute system consisting of the colloid-polymer micelles and the free polymer chains. Equation (3) can still be used except one has to replace f_1 by $f_1 + \alpha\omega f_2$ and ρ_2' by $(1 - \alpha)\rho_2'$. The upper solid line in Fig. 4 is a fit to the linear function $3.7(1 - 27.4\rho_2')$. The effective slope of the linear function is $27.4/(1 - \alpha_0) = 34.7$, which is a factor of 0.68 smaller than that for the colloid/PEP mixture. The smaller slope in $P(\rho_2')/b_{11}(0)$ indicates that the polymer adsorption greatly reduces the depletion attraction between the colloidal particles. However, there is still some attraction between the colloid-polymer micelles due to the un-adsorbed

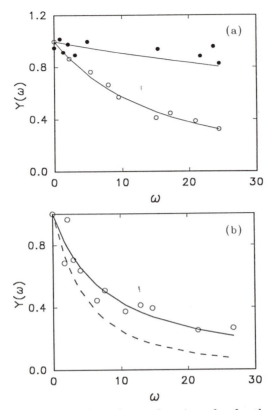

Figure 3. The measured $Y(\omega, \alpha_0)$ as a function of ω for the colloidal mixtures with PEP (closed circles in 3a), amine-PEP (open circles in 3a), and zwitterion-PEP (open circles in 3b). The solid curves are the least-square fits to the data points (see text).

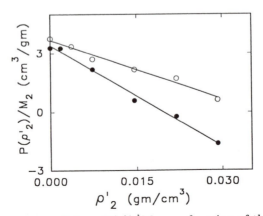

Figure 4. The measured slope $P(\rho_2')/M_1$ as a function of the polymer concentration ρ_2' for PEP (closed circles) and for amine-PEP (open circles). The solid lines are the linear fits to the data points.

polymer molecules in the solution. This is shown by the decreasing trend of $P(\rho_2')$ with increasing polymer concentration. The stabilization of our colloidal mixture with the end-functionalized polymers is also observed in a simple phase study. Three colloidal mixtures were prepared with different polymers: PEP, amine-PEP, and zwitterion-PEP. The three samples have the same colloid concentration (9% by weight) and the polymer concentration (5.3% by weight). The three polymers also have approximately the same molecular weight ($M_w \simeq 25,000$). It was observed that the sample with PEP phase separated, as is expected for the depletion effect. The other two samples were clear, and no sign of phase separation was observed.

Conclusion

A static light scattering method is used to study the adsorption of end-functionalized polymers on colloidal spheres. The small-angle light scattering intensity from the colloid-polymer mixture is calculated using a virial expansion method for binary scatterers. With the obtained formula one can measure the polymer partition coefficient α, which is the probability for a single polymer chain to anchor to the colloidal surface. With this knowledge, we compute the adsorption energy ϵ between the end-functional group and the colloidal surface. The scattering scheme also shows how the light scattering data provides information about changes of the microscopic interactions in the colloid-polymer mixture.

In the experiment the monodispersed hydrogenated-polyisoprene (PEP) and its single-end-functionalized derivatives are used to modify the interaction between the polymer and the colloidal surface. Measurements of the scattering intensity reveal that only a fraction of the end-functionalized polymers are adsorbed on the colloidal surface. The results for the end-functionalized polymers are compared with those for the un-functionalized polymer. It is found that the interaction between the colloid and the un-functionalized polymer is repulsive, which introduces a depletion-attraction between the colloidal particles. The functional end-groups are found to interact attractively with the polar cores of the colloidal particles. The adsorption energy between the functional group and the colloidal surface is estimated to be $\sim 4k_BT$. The presence of the adsorbed polymer on the colloidal surfaces greatly reduces the depletion attraction between the colloidal particles, and therefore, enhances the stability of the colloid-polymer mixture. The experiment is of interest to observe the microscopic interaction between colloid and polymer and to see how it responds to the incorporation of functional group on the polymers. We expect our scattering technique to be useful for measuring the adsorption of polymer molecules in colloidal suspensions.

Acknowledgments

We have benefited from illuminating discussions and correspondence with T. A. Witten at all stages of this work. One of us (P.T) acknowledges the Donors of The Petroleum Research Fund, administered by the American Chemical Society, for the partial support of this work. He also thanks S. Milner and Z-G. Wang for useful discussions.

Literature Cited

[1] M. A. Cohen-Stuart, T. Cosgrove and B. Vincent, Adv. Colloid Interface Sci., **24**, 143 (1986).

[2] D. H. Napper, *Polymeric Stabilization of Colloidal Dispersions*, (Academic, New York, 1983).

[3] see, e.g., G. P. Van der Beer and M. A. Cohen Stuart, J. Phys. France, **49**, 1449 (1988).

[4] A. Vrij, Pure Appl. Chem., **48**, 471 (1976).

[5] S. Asakura and F. Oosawa, J. Chem. Phys., **22**, 1255 (1954).

[6] P. Tong, T. A.Witten, J. S. Huang, and L. Fetters, J. Phys. (France), **51**, 2813 (1990).

[7] N. W. Ashcroft and D. C. Langreth, Phys. Rev. **156**, 685 (1967).

[8] P-G. de Gennes, *Scaling Concepts in Polymer Physics* (Cornell, Ithaca, 1979).

[9] E. Eisenriegler, K. Kremer, and K. Binder, J. Chem. Phys., **77**, 6296 (1982).

[10] A. L. R. Bug, M. E. Cates, S. A. Safran, and T. A. Witten, J. Chem. Phys., **87**, 1824 (1987).

[11] Z-G. Wang, Langmuir, **6**, 928 (1990).

[12] I. Markovic, R. H. Ottewill, D. J. Cebula, I. Field, and J. F. Marsh, Colloid and Polymer Sci., **262**, 648 (1984); I. Markovic and R. H. Ottewill, Colloid and Polymer Sci., **264**, 65 (1986).

[13] I. Markovic and R. H. Ottewill, Colloid and Polymer Sci., **264**, 454 (1986); Colloids and Surfaces, **24**, 69 (1987).

[14] J. Mays, N. Hadjichristidis, and L. Fetters, Macromolecules, **17**, 2723 (1984); **22**, 921 (1989).

[15] N. S. Davidson, L. J. Fetters, W. G. Funk, W. W. Graessley, and N. Hadjichristidis, Macromolecules, **21**, 112 (1988); **20**, 2614 (1987).

[16] H. J. Taunton, C. Toprakcioglu, L. J. Fetters and J. Klein, Nature **332** 712 (1988).

[17] C. Ligoure and L. Leibler, J. Phys. France, **51**, 1313 (1990).

[18] H. De Hek and A. Vrij, J. Colloid Interface Sci., **84**, 409 (1981).

RECEIVED February 23, 1993

Chapter 10

Adsorption of End-Functionalized and Cyclic Polymers

T. Cosgrove[1], R. D. C. Richards[1,3], J. A. Semlyen[2], and J. R. P. Webster[1,4]

[1]School of Chemistry, University of Bristol, Cantock's Close, Bristol
BS8 1TS, United Kingdom
[2]Department of Chemistry, University of York, Heslington, York YO1 5DD,
United Kingdom

The effect of the end groups of polymer chains on adsorption at the solid-solution interface has been studied. The results show that changes in the end group moiety can be sufficient to enable one polymer to displace an otherwise identical polymer from the interface. The preferential adsorption of end groups is also shown by comparing the adsorption isotherms of linear and cyclic polymers.

The structure of a linear homopolymer chain is such that the main chain segments can never be identical, in the chemical sense, to the end segments; in reality the only true homopolymers must be cyclic. The effect of end group functionality on the adsorption properties of polymers has been seen in selective deuteration studies in polymer melts [1], where surface enrichment of the deuterated polymer was found, and in a systematic series of data on the adsorption of end-functionalised polydimethylsiloxane polymers on silica [2]. In this latter study, greater adsorbed amounts were found when the end segments had a high affinity for the surface. In another series of experiments comparing the adsorption of cyclic and linear polymethylsiloxane polymers [3], it was found that for low molecular weight chains of the same number of monomers, the cyclic polymers adsorbed to a considerably greater extent than the linears. In an attempt to rationalise this data, theoretical calculations using a self consistent mean-field theory [4] were undertaken. It was found that although the model reproduced the

---

[3]Current address: Department of Chemistry, University of Hull, Hull, United Kingdom
[4]Current address: Rutherford and Appleton Labs, Chilton, Didcot, Oxon, United Kingdom

sense of the adsorption, the absolute differences between the cyclic and linear polymers were rather small. In an attempt to reconcile this difference a very unfavourable adsorption energy for the end groups was proposed [5].

In this study we present data on two systems, where both these phenomena are reproduced. The difference in solvency of cyclic and linear polymers is also discussed. The results are compared with calculations based on the Scheutjens-Fleer theory [4] and by Monte Carlo simulations [6].

Experimental

Materials: The poly(ethylene oxide) samples used were commercial samples obtained from Shell U.K. and their structure and molecular weights are shown in Table 1.

Table 1 Poly(ethylene oxide) Characterisation

Nomenclature+	M_W	Structure
PEO750	750	$CH3-[CH2-CH2-O]_{17}-H$
PEG800	800	$OH-[CH2-CH2-O]_{18}-H$

The polystyrene samples were prepared by following the scheme given by Guiser and Höcker [7] except that tetrahydrofuran [THF] was used as the solvent and dimethyldichlorosilane as the coupling agent [8]. Half of the living polymer mixture was added dropwise and simultaneously with a 2% solution of the coupling agent to a large bath of THF. This ensured that the concentrations of both living polymer and coupler were always low. The other half of the living polymer sample was terminated with methanol to give a comparable molecular weight linear fraction. The cyclic material was separated by fractional precipitation from THF with methanol. This precipitation was monitored by GPC. The cyclic polymer was analysed by ^{29}Si NMR and by photon correlation spectroscopy to identify it as cyclic. These latter results were in agreement with those published by other authors [9]. The characterisation of the samples is given in Table 2.

Table 2 Linear [Lx] and cyclic [Cx] polystyrene samples: x is the number of monomers

Linear	M_n	Cyclic	M_n
L41	4,300	C41	4,100
L140	14,500	C123	12,800
L182	19,000	C193	20,000

The carbon substrate was a graphitized carbon black, Graphon [Cabot U.K.], with a   specific surface area of 80 m^2/g.   The γ-alumina sample was obtained from Prest [Grenoble France] and had a specific surface area estimated at 200m^2/g. However this sample consisted of aggregates of approximately 1 μm in size which were made up of smaller particles of the order of 9nm diameter. The area available for polymer adsorption was therefore considerably less than the nominal nitrogen BET area quoted above. The tetrachloromethane and cyclohexane were obtained from the Aldrich Co. [U.K.] and were spectroscopic grade.

Methods:

The adsorption isotherms for the poly(ethylene oxide) samples on Graphon from tetrachloromethane were obtained by using high resolution narrow-bandwidth NMR. Because of the settling of the dispersion and the fact that for low molecular weight polymers the NMR signal linewidth is substantially broadened on adsorption the signal that is measured in this way can be mainly associated with that from the bulk solution. This then gives an estimate of the equilibrium solution concentration [10].   Figure 1 shows the NMR spectrum from the PEO750 polymer showing that both the CH$_3$ and the CH$_2$ peaks can be resolved in the dispersion. By a suitable calibration procedure with polymer solutions of known concentration, the peak integral can be used to measure the equilibrium polymer concentration in the dispersion. From the known molecular structure of the PEO750 polymer, the concentration of the two different polymers in a mixture can be also be found by ratioing the methyl and methylene peaks as follows.-

$$\frac{[PEG800]}{[PEG750]} = \frac{[\int CH_2 - \alpha \int CH_3]}{\alpha \int CH_3} \qquad [1]$$

where α^{-1} is the ratio of the number of protons in the methyl group to the total number of methylene protons in the PEO750 chain [~ 23].

The adsorption isotherms for the polystyrene samples were obtained by analysing the supernatant solution after adsorption and centrifugation by monitoring the UV absorption band at 240nm.

Results and Discussion:

[1] Functionalized homopolymers: Figure 2 shows the   adsorption isotherms obtained for   the two poly(ethylene oxide) polymers adsorbed on Graphon from solution in tetrachloromethane. The isotherms are not of the high affinity type due partly to the polydispersity of the polymers [Mw/Mn ~1.5-2.0] and their relatively low molecular weight. The values of the adsorbed amount are not given on an absolute adsorbance scale as the dispersion surface area is not known accurately, and a small correction has to be made for the signal that could arise from mobile segments in the adsorbed layer [11]. The results show that the PEO750 polymer

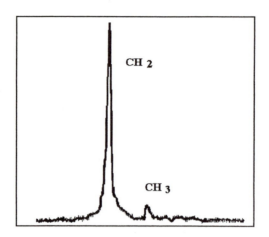

Figure 1 High resolution 100 Mhz spectrum of PEO750 in a
Graphon/carbon tetrachloride dispersion, showing both the methylene and
methyl protons.

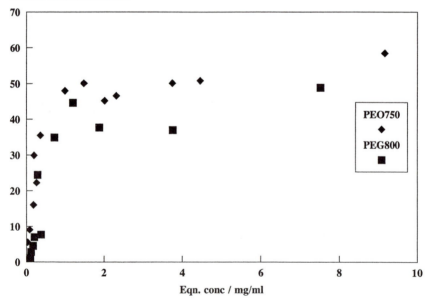

Figure 2   Adsorption isotherms for the PEO750 [◆] and the PEG800 [■]
polymers on Graphon from tetrachloromethane.

adsorbs more than the PEG80O polymer even though it has a marginally lower degree of polymerisation [Table 1]. This small difference in adsorption which is just outside the experimental error may be understood in terms of the different surface affinities of the polymer end groups. The carbon surface is essentially hydrophobic and thus its interaction with the hydroxyl end groups is therefore likely to be rather weak. However the methyl end group of the PEO750 will be more strongly attracted to the surface. This suggests that the tail volume fraction profiles for the two polymers may be somewhat different, the PEO750 losing some of its' end segments by preferential adsorption at the surface. In order to assess the likely effect of this specific adsorption, we have carried out a calculation of the adsorption isotherms for two polymers, modelled as block copolymers, using the Scheutjens Fleer theory [12].

Table 3 Parameters used in the theoretical calculations for the polyethers

Parameter	OH[A]	CH_3 and $[CH_2CH_2O]$ [B]
χ	0.45	0.45
χ_S	0.0	0.6

The thermodynamic parameters, χ [Flory Huggins parameter] and χ_S [The net surface adsorption energy] are given in Table 3, for two polymers of structure A1B18A1 and A1B19. It is assumed that the χ_S for the methyl and ethylene oxide segments will be the same and no scaling of the size of the segments to the lattice has been undertaken. The resulting isotherms are given in Figure 3. The small preferential adsorption of the methyl end segment does lead to a slightly higher adsorbed amount for the PEO750 as seen in the experiments. The theoretical isotherms are also of low affinity and do not show a clear plateau even at an equilibrium concentration of 10,000 ppm. For the experimental isotherms, it is not possible to make any firm conclusions on the slope of the isotherm beyond the knee, at least within the estimated error of the measurements [10%]; the solid lines shown in Figure 2 are simply to guide the eye. Figure 4 shows volume fraction profiles for the tail segments for the two polymers calculated by using the Monte Carlo method with a fixed adsorbed amount, θ of 0.8 for both samples [6]. The same thermodynamic parameters were used as in Table 3. The profiles show that given a small differential end group adsorption energy, the tail volume fraction can be suppressed and the sum of the train and loop populations augmented. The increased adsorbed amount for PEO750 will offset this decrease in the tail fraction to some extent, but it is likely that the excess adsorbed amount in the experimental system is accommodated mainly by more segments in loops and to a lesser extent in trains. For further comparison a B20 homopolymer where all the segments have

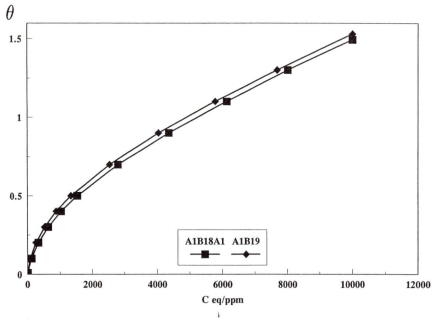

Figure 3   Self   consistent   mean-field   calculations   for   the   adsorption
isotherms for two block copolymers of structures A1B18A1 [■] and A1B19[
◆]. The values used for the χ parameters are given in Table 3.

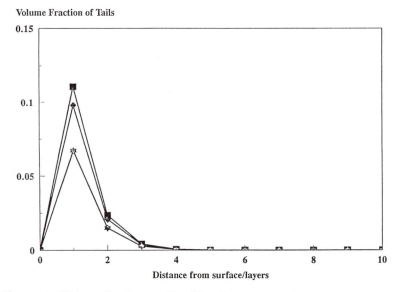

Figure 4   Volume fraction profiles for tail segments using the Monte Carlo
simulation method. $\theta = 0.8$. The other parameters are given in Table 3.
A1B18A1[■] and A1B19[◆] and B20 [✶].

the same surface affinity is also shown in Figure 4. Here it is seen that the number of tails segments is substantially suppressed compared to the other polymers given that the adsorbed amount is the same. Given the differences in the structure of the adsorbed layer it would suggest that the PEO750 polymer might displace the PEG800 polymer from the surface and Figure 5 illustrates such a study. In this experiment initially, the PEG800 was adsorbed from a dilute solution on the rising part of the adsorption isotherm [Figure 2]. To this sample, an increasing amount of PEO750 was added. The equilibrium concentration of both species was monitored from the NMR spectrum as explained above. The results shown in Figure 5 indicate that the PEG800 is completely displaced by the PEO750 given a sufficient solution concentration of the latter [$\approx$ 1.5 mg/ml]. At low equilibrium concentrations, corresponding to the rising part of the adsorption isotherm, both polymers are adsorbed in a mixed monolayer. This behaviour is rather similar to the more familiar solvent displacement isotherms [13]. The critical displacement concentration however is not so strongly marked in this example.

[2] Linear and Cyclic polymer adsorption: Figure 6 shows the adsorption isotherm   values for the cyclic polystyrenes adsorbed on γ alumina from cyclohexane at room temperature as a function of molecular weight. At this temperature, both the linear and cyclic polymers are below their theta temperatures and the adsorption isotherms do not show a clear plateau, but a change in slope, similar to the theoretical isotherms shown in Figure 3 for the poly(ethylene oxide) system. The points selected on the isotherms to obtain Figure 6 were at an equilibrium concentration of 650ppm where there was a substantial change in curvature. For all three samples, the adsorbed amount for the cyclic polymer exceeds that of the linear, though the data appear to converge at the higher molecular weight end. This compares favourably with the data for polymethylsiloxane [3] and polydimethylsiloxane [2] polymers adsorbed on both silica and alumina which show a similar behaviour. A major difference however in this case is that the polymers are in a worse than θ solvent. The θ temperatures for the linear and cyclic polymers are 309 and 301K respectively. At 298 the linear is therefore in a worse solvency situation which should enhance its adsorption over the linear. This may become more important with higher molecular weight polymers and contribute to the convergence of the two sets of data. Figure 7 shows a comparable calculation for this case. The χ values at room temperature were obtained by scaling the differences from the respective theta temperatures. This gave 0.52 for the linear and 0.51 for the cyclic and both polymers were thus in worse that θ conditions. χ_S was taken as 1.0 and an equilibrium concentration of polymer of 650 ppm. The plateau values for both the linear and cyclic polymers are very similar, and a crossover of the chain length dependencies at a value of r of 65 can be seen. Given that the persistence length of polystyrene is about 6 monomers would place the cross-over in the experimental system at $\approx$ 40,000. The experimental data in Figure 6 are consistent with this prediction. The effect of

Adsorbed amount mg/ml

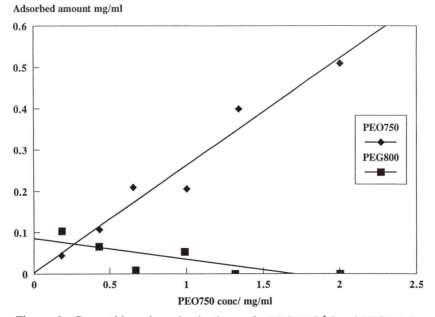

Figure 5   Competitive adsorption isotherms for PEO750[◆] and PEG800 [
■] on Graphon in tetrachloromethane.

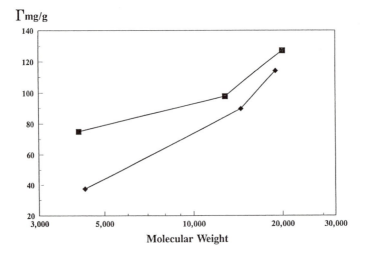

Figure 6   Adsorbed amount against M_n for linear [◆] and cyclic [■]
polystyrene adsorbed on γ-alumina from cyclohexane at room temperature.
The equilibrium polymer concentration was 650ppm.

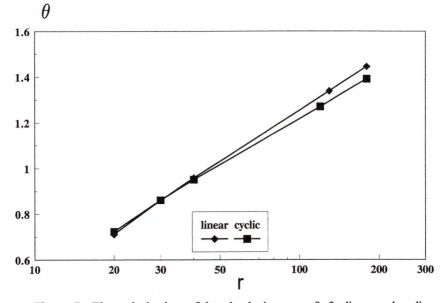

Figure 7   Theoretical values of the adsorbed amount, θ, for linear and cyclic polystyrenes as a function of chain length, r. χ LINEAR = 0.5 2 , χCYCLIC = 0.51 and χ_S = 1.0. The equilibrium polymer concentration was taken as 650 ppm. linear [♦] cyclic [■].

the adsorption of end groups for the linear polystyrene chain are complicated with one end favouring adsorption [OH] on the γ-alumina whilst the other, being derived from the initiator [Sodium Napthenide 8], is likely not to be preferentially adsorbed. As with the earlier data for the adsorption of linear and cyclic polymers [2,3] it would appear that the linear polystyrene chain has a marginally smaller enthalpic preference for the surface [through the end groups] than the cyclic.

Conclusions

Preferential adsorption effects have been found for different end-functionalized polymers and in comparing linear and cyclic polymers of the same monomers. These effects are much stronger than would have been anticipated and may have implications in studies where differently ended polymers are being compared.

Acknowledgements: RDCR would like to acknowledge the SERC and DOW Chemical for the provision of a CASE award. JRPW would like to thank EXXON chemical for the provision of a Post Doctoral Fellowship. The SERC computational science initiative is acknowledged for the provision of the MEIKO/Sun supercomputer system. The authors would also like to thank Dr. T. Kendrick [DOW] for his continued support of the adsorption project.

References

1] Jones, R. Private Communication
2] Patel. A.. D. Phil Thesis University of York [1991]
3] Cosgrove, T.; Prestidge, C.A.; Vincent, B. J. Chem. Soc. Faraday Trans. **1990,** 86, 1377
4] van Lent, .; Scheutjens, J.M.H.M.; Cosgrove, T. Macromolecules **1987,**20, 366.
5] Cosgrove, T.; Patel, A.; Semlyen, J.A.; Scheutjens, J.M.H.M.[Submitted to Macromolecules]
6] Cosgrove, T.; . Finch, N.A.; . Webster, J.R.P. Macromolecules **1990,** 23, 3353
7] Guiser, D.; Höcker, H. Macromolecules **1980,**13, 653
8] Richards, R.D.C., D. Phil. Thesis University of York [1988]
9] Hadziioannou, G. Cotts, P.M.; ten Brinke, G.; Han, C.C.; Lutz, P.; Strazielle, C.; Remp, C.; Kovacs, A.J. Macromolecules **1987,**20, 493
10] Cohen Stuart, M.; Cosgrove, T.; Vincent, B. Adv. in Coll. and Int. Sci. **1986** 24, 142.
11] Cosgrove, T.; Fergie-Woods, J. Colloids and Surfaces. **1987,**25, 91
12] Evers, O.A.; Fleer, G.J.; Scheutjens, J.M.H.M. J. Chem. Soc. Faraday Trans. 990 ,86, 1333
13] Van der Beek, G.; Cohen Stuart, M.; Cosgrove, T. Langmuir **1991,**7, 327

RECEIVED February 23, 1993

Chapter 11

Application of Electro-optics To Investigate the Electrical and Hydrodynamic Properties of Colloid–Polymer Surface Layers

Raphaël Varoqui[1], Tsetska Radeva[2], Ivana Petkanchin[2], and J. Widmaier[1]

[1]Institut Charles Sadron (CRM-EAHP), Centre National de la Recherche Scientifique–ULP Strasbourg, 6 rue Boussingault, 67083 Strasbourg Cédex, France
[2]Institute of Physical Chemistry, Bulgarian Academy of Science, 1040 Sofia, Bulgaria

Light scattering in the presence of an electrical field (electro-optics) was recorded on aqueous suspensions of anisometric shaped colloidal particles having an adsorbed polymer layer. The light scattered in the presence of an electrical-field can be related to the electrical polarizability of the interface. It is shown that the adsorption of polyacrylamide (a water soluble neutral polymer) on β ferric hydrous oxide particles decreases strongly the polarizability of the counterion cloud around the colloid. This was also confirmed by electrophoretic mobility measurements. These results are explained by the attachment of polymer to hydroxyls of the colloid surface. Hydrogen bonds among the colloid and the polymer units results in a decrease of the net surface charge. Furthermore, the thick and dense polymer layer may depress the dielectric constant of the water in the interfacial region. The size of the adsorbed layer was determined from : (i) the time decay of the electro-optical effect which gives the rotary diffusion coefficient of the colloid-polymer complex ; (ii) quasi-elastic light scattering which yields the translational diffusion coefficient. An average hydrodynamic thickness of the polymer layer was determined from the decrease of the diffusion coefficients in presence of polymer. The molecular weight dependence of the layer thickness was analyzed using De Genne's theory in which a power law is derived for the concentration profile of monomers.

In electro-optical measurements, one determines essentially an electrical dipole moment which is related to the number and the spatial distribution of mobile ions in the vicinity of a colloid surface. This technique - light scattering in the presence of an electrical field - was abundantly used in the investigation of the electrical properties of colloids dispersed in water. Extensive monographs on colloid electro-optics were published (1-2). In recent years, the effect of water soluble polymers on the behaviour of hydrosols (especially colloids of mineral or metallic origin) began to raise interest. Polymers partly because of their large

0097–6156/93/0532–0121$06.00/0

size, may act as efficient flocculants and/or dispersants of colloidal particles which are of great technical importance (*3-4*). This has prompted during the last decade a copious amount of experimental and theoretical work in order to elucidate the features of the conformations of polymer chains confined to a solid/liquid interface. Polymers at interfaces are generally viewed as a succession of monomer sequences in "contact" with the surface and large loops which extend into the solution phase over distances comparable to the polymer size. The basic concepts such as the average layer thickness, average concentration profile were mainly developed when the monomers interact with the surface via "weak" forces which is typical for dispersions of colloids interacting with non-charged polymers in organic liquids (organosols). In aqueous environment, the "surface" or phase boundary between the colloid and the suspending medium is charged, the adsorbed polymer might be either neutral or bear charges of sign opposite to the colloid charge. The concern of this work is to show that in using the electro-optic technique, it is possible to reach information about the electrical properties of the colloid/polymer complex as well as on the structure of the polymer in the adsorbed state. Two aspects should be pointed out:

-At present, the extend to which the adsorption of neutral polymers do affect the counterion layer of the colloid is not exactly known. Polymer layers in depressing the counterion activity could release electrostatic repulsions among colloids.

-In the case of polyelectrolyte adsorption, the electrical dipole comes from both colloid and polyelectrolyte counterion charges. Electro-optical measurements are likely to provide information on the structure of charged polymers in the adsorbed state.

In this work, we address to the first point : our objectives were to explore the modifications of the electrical double layer and hydrodynamic properties of a β ferric hydrous oxide colloid in aqueous media, interacting with non-hydrolyzed polyacrylamide, a neutral polymer. The β ferric hydrous oxide particles are ellipsoidal shaped particles and the electrical double layer features can be determined through the electrical polarizability of the ellipse. Electro-optical determinations were complemented by measuring the electrophoretic mobility at various polymer coverage.

The electrical moment is measured under steady-state light scattering. When the electrical field is switched off, the electro-optical signal decays because particles tends to become randomly oriented. The width of the exponential decay is related to the rotary diffusion coefficient of the particle. It is then possible from the rotary diffusion coefficient of the colloid/polymer ensemble, to derive an effective hydrodynamic thickness of the polymer layer. This parameter was compared to the hydrodynamic thickness obtained from the translational diffusion coefficient by photon correlation spectroscopy. A large range of molecular weight was investigated to establish the dependence of the hydrodynamic thickness with the chain length. Theoretical studies have amply confirmed that this relation is closely connected with the structural features of the polymer layer (*5-6*).

Experimental Section

Materials. Colloid Particles. The ferric hydrous oxide sol was prepared by acid hydrolysis of 0.019 M $FeCl_3$ solutions containing 10^{-3} M HCl over a period of three weeks at room temperature. This yielded the β form of the oxide-hydroxide (*7-8*). Free Fe^{3+} ions were removed by repeated centrifugation. In some instances, dialysis of the sol against deionized water was also done to obtain the sol free of Fe^{3+} ions. The structure of βFeOOH is of hollandite type. The method of preparation gives ellipsoidal shaped particles uniform in shape and of narrow size distribution - Figure 1. The morphology was studied extensively and needs no further discussion (8-9). By electron microscopy we found the dimensions a and b of the major and minor axis of the prolate ellipsoid to be respectively 285 and 72 nm (axial ratio 4). Some determination of the electrokinetic and thermodynamic properties of ferric hydrous oxides have been reported (10-11). Surface charges results from interactions of protons with surface hydroxyl groups to give in acidic medium $FeOOH_2^+$ groups, whereas in basic medium, protons are released and FeO^- groups are formed.

We have determined the ζ potential at different pH - Figure 2. The isoelectric point is found by interpolation to be near pH7 which agrees well with former determination of the pcz on $\alpha FeOOH$ sols of goethite structure (12). It is important to note that the electrokinetic charge changes in a narrow pH range. At pH 5.5 at which most of our data were recorded, the net (positive) charge does not differ by more than 15% from its maximum value recorded at pH 4.3.

Polymer. Polyacrylamide was prepared by radical polymerization of acrylamide in aqueous media according by known techniques (13). The polymer was then fractionated in a water methanol mixture as previously described (14). Fractions with molecular weight in the range of 3.03×10^5 to 1.9×10^6 were obtained. Fractions were analyzed for polydispersity by GPC. Polydispersity was found to be between 1.1 and 1.2. The polydispersity is slighlty larger for the high molecular-weights.

For the determination of the adsorbed amount a radioactive polyacrylamide of molecular-weight 1.2×10^6 was prepared with specific radioactivity 0.05 m Ci/g (15). The radioactive label was ^3H :

$$[CH_2\text{-}CH(CONH_2)]_m[CH_2\text{-}CHCHOH^3H]_n, \quad m/n = 10^3 \tag{1}$$

Methods. Adsorption Isotherm Determination. The adsorption isotherm was established by mixing aliquots of a stock solution of radioactive polyacrylamide $M_w = 1.2 \times 10^6$ of known concentration with a given volume of $\beta FeOOH$ suspensions with a final concentration in colloid of 4×10^{-2} g/l. After pH adjustment, and gentle tumbling at 25°C during 24 Hrs, the mixture was centrifuged and an aliquot of the supernatant was counted for radioactiviy. The adsorbed amount C_S expressed in g per g colloid was determined by following relation :

$$C_S = P^{-1}[M - R^*/R^*_{sp}(V\text{-}P/d)] \tag{2}$$

M is the weight of polymer in the mixture, P the weight of the colloid, d the density (4.28 g/cm³), V the volume of the mixture and R^*, R^*_{sp} are respectively the radioactivity per unit volume of the supernatant and the specific radioactivity of the polymer.

Electro-Optical Measurements. Steady State Electro-Optical Effect. The latter is defined as :

$$\alpha_S = (I_E\text{-}I_0)/I_0 \tag{3}$$

I_E and I_0 are the intensity of scattered light when an electrical field of strength E is applied and at zero field. Light scattering was recorded at an angle of $\pi/2$ with respect to the electrical field of strength 180 V cm⁻¹ and frequency 1 KHz (all technical details are reported in ref.(16). In the case of cylinder symmetrical particles with axis of length a and b, at low degrees of orientation, and in the Rayleigh-Debye-Gans approximation, α_S is given by :

Figure 1.:   Electron micrograph of β ferric hydrous oxide particles.

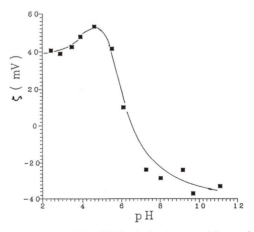

Figure 2.:   Zeta potential of β ferric hydrous oxide particles vs pH.

$$\alpha_S = \frac{A(Ka,Kb)}{I_0(Ka,Kb)} (\mu^2+\delta)E^2 \qquad (4)$$

$\mu = p'/k_BT$, p' being the permanent dipole along the symmetry axis of the particle, $\delta = (\gamma_1-\gamma_2)/k_BT$, γ_1 and γ_2 being the electrical polarizabilities along the symmetry and transverse axis respectively (responsable for the induced electrical moment, $p_{ind}=\gamma E$), K is the magnitude of the wave vector. The functions $A(Ka,Kb)$ and $I_0(Ka,Kb)$, are represented in ref.(*17*).

Decay of the Electro-Optical Signal. After switching off the electrical field the electro-optical signal α_t decays exponentially with time (*2*) :

$$\alpha_t = \alpha_S \, exp(-6 \, D_r t) \qquad (5)$$

For prolate ellipsoids, the rotary diffusion coefficient D_r is given by Perrin equation (*18*) :

$$D_r = \frac{k_BT \, p^2}{4\eta v(p^4-1)} \left[-1 + \frac{2p^2-1}{2p \, \sqrt{p^2-1}} \, Ln \, \frac{p + \sqrt{p^2-1}}{p- \sqrt{p^2-1}} \right] \qquad (6)$$

p is the axial ratio a/b, η the viscosity of the suspending medium and $v = 6^{-1}\pi ab^2$ is the volume of the particle. In presence of an adsorbed polymer layer, from the change in the axial ratio p determined through D_r, we define the effective hydrodynamic thickness L_r of the polymer by :

$$p = (a+2L_r)/(b+2L_r) \qquad (7)$$

Translational Diffusion Coefficient. A Malvern Zeta Sizer was used for the determination of the translational diffusion coefficient D_t through photon correlation spectroscopy with the Az4 cell and light scattered at an angle of $\pi/2$ with respect to the incident beam. The translational diffusion coefficient D_t is related to the dimensions by following equation (*18*) :

$$D_t = \frac{k_BT}{3\pi\eta b} \frac{Ln \, (p+\sqrt{p^2-1})}{\sqrt{1-p^2}} \qquad (8)$$

From the change of p in presence of polymer the effective hydrodynamic thickness L_t is obtained by equation 7 replacing L_r by L_t.
The Stokes radius R_h ($R_h = k_BT/6\pi\eta \, D_t$) of the bare colloid measured by photon spectroscopy at several pH values is reported in Table I.

Table I. Hydrodynamic Radius at Different pH of β Ferric Hydrous Oxide Colloid Particles

pH :	2.4	2.9	3.5	3.9	4.6	5.5	7.3	8.0	9.1	9.7	11.1	
R_h(nm) :		68	73	70	77	75	77	73	77	87	75	70

Values are scattered over ± 7% but no systematic variation with the pH is observed. The value 77 at pH 5.5 is near to the average value (75.5 nm). This proves indirectly that particles are not aggregated at that pH. From the average value and applying equation 8, we obtain for a and b, 322 nm and 80.4 nm respectively. Since hydrodynamics cannot resolve separately for a and b via equation 8, p was taken equal to 4 as in the electron microscopic observation. From the rotary diffusion coefficient, the dimensions of the bare colloid were found to be 345 and 86 nm. The dimensions determined by both techniques which do not differ by more then 6% are substantially larger than the values determined by electron microscopy. This discrepancy may find its origin in size polydispersity. Indeed, the orientation relaxation time as well as the mean translational time measured by light scattering are very sensitive to the largest particles present in the suspension (D_r and D_t are respectively $z+1$ and z averages) (16,19). Since the colloid sample is not completely devoid of any (though tiny) size polydispersity, the hydrodynamic data may reflect somewhat more the dimensions of the largest particles.

The concentration of the suspension was 4×10^{-3} g/l in the electro-optic technique and 4×10^{-2} g/l in the Malvern technique. The pH, unless otherwise specified was 5.5 and the ionic strength 10^{-5}. Each sample suspension was sonicated at 22 KHz for 3 min before the measurement was started.

Electrophoretic Mobilities. The electrophoretic mobilities reported in Figures 4 and 6 were measured using a Rank Brother apparatus Mark II and flat quartz cell at 25°C. At least 20 particles were timed at both stationary layers. The error is within 5 to 10%.

Results and Discussion

Adsorption Isotherm. The adsorption of PAM (Mw = 1.2×10^6) was measured on β-FeOOH at pH 4 and 5.5. Adsorption increases steeply at very low polymer bulk concentration - Figure 3. The plateau values of the adsorbed amounts are C_S = 0.10 g PAM/g colloid at pH 4 and C_S = 0.40 g PAM/g colloid at pH 5.5. The pH dependence of the plateau value is consistent with previous observation in a work to elucidate the adsorption mechanism of this polymer on metal oxide surfaces. The amount of adsorbed polymer was found to vary in exact proportions with the number of residual M(OH) metal hydroxide on the surface and is higher at smaller surface charge (20). It was concluded that the polymer interacts with the carbonyl of the amide $CONH_2$ moiety and surface hydroxyls through hydrogen bonding.

Electro-Optical Properties. In Figure 4 are represented for molecular weight 1.2×10^6, the electro-optical effect α_S, the electrophoretic mobility U_E and the adsorbed amount C_S as a function of the concentration C_0(g/l suspension) of polymer added to the suspension. One observes that α_S and U_E scarcely change up to C_S 4.10^{-3} g/g after which both quantities decrease sharply. The surface concentration C_S reported in Figure 4 was calculated from the isotherm determined at pH 5 (Figure 3) and from the amount of polymer added to the suspension. The decrease of α_S and U_E corresponds to a steep increase in adsorption. It should however be noted that the α_S and U_E variation takes place in a concentration range where the amount of polymer at the interface is still far from its saturation value. For instance, U_E becomes zero for C_S 4.5×10^{-2} g/g which is about a tenth of the plateau value. We infer therefore that small amounts of neutral polymer in the adsorbed state modify considerably the electrical and electrokinetic properties of the colloid. U_E decreases monotically, whereas α_S exhibits a slight hump. As observed in Figures 5 and 6, where α_S and U_E are plotted for different molecular weights, the effect of the chain length is not large.

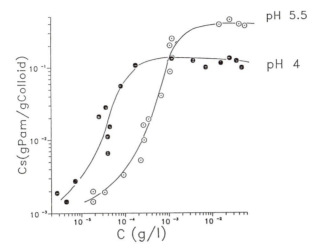

Figure 3.: Adsorption isotherms of polyacrylamide of molecular weight 1.2×10^6 at pH 4 (●) and pH 5.5 (O).

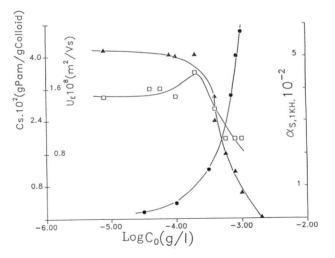

Figure 4.: Surface concentration C_s vs log of initial concentration C_0 (●) ; electrophoretic mobility U_E vs log C_0 (▲) ; electro-optical effect α_s vs log C_0 (▢). Molecular weight Mw = 1.2×10^6.

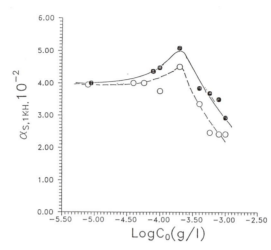

Figure 5.:   Electro-optical effect α_s vs log of initial concentration C_0 (●)
Mw = 3.03×10^5 ; (O) Mw = 1.2×10^6.

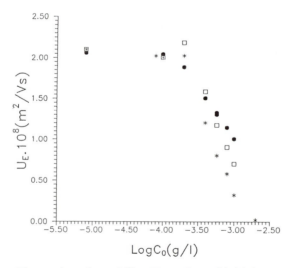

Figure 6.:   Electrophoretic mobility U_E vs log of initial concentration C_0
for different molecular weights ; (*) Mw = 1.2×10^6 ; (◻) Mw
= 5.93×10^5; (●) Mw = 3.03×10^5.

From analysis of the α_S dependency on field frequency (dispersion curves) on different colloids, it has been shown that the orientation of particles in the kHz range originates solely from the induced dipole moment *(2,16,21)* and it is generally accepted that the induced dipole moment or electric polarizability γ is due to the movement of ions in the diffuse part of the electrical double layers *(16,21-23)*. The prefactor $A(Ka,Kb)/I_0(Ka,Kb)$ could also depend on C_S. From model calculation, it appears that this factor should increase for positive Δa and Δb increments. Since no significant change in I_0 (the light scattered at zero field) or α_∞ (the electro-optical effect at full particle orientation) was observed in the presence of polymer, we believe that the optical factor is also constant. We shall now discuss a mechanism which could possibly explain the observed effects.

i) At low surface coverage ($C_S < 4.10^{-3}$ g/g), the constancy of α_S and U_E shows that the diffuse part of the electrical double layer is not influenced by the adsorbed polymer. This is possible at low polymer concentrations when the polymer is in a flat conformation on the surface. At $C_S > 4.10^{-3}$ g/g the slope of the adsorption isotherm becomes large, which may be associated with conformational changes in the polymer layer when loops are built. This might also be the reason for the considerable changes in the diffuse part of the electrical double layer and in particular in α_S and U_E. Similar behaviour of α_S and U_E over a narrow range of polymer concentrations has already been observed *(24,25)*. A steep increase of the hydrodynamic layer thickness (Figure 5) and the stabilization of the suspension at $C_S > 4.10^{-3}$ g/g (suspension stability was checked by addition of 10^{-1} M NaCl) are in line with this explanation.

The decrease of α_S could be explained by the decrease of the surface conductivity due to the presence of neutral polymer molecules in the diffuse part of the electrical double layer *(25)*. Another reason for the decrease of the polarizability γ could be the decreased dielectric permitivity ε of the solvent. From theoretical considerations we expect γ to vary as $\varepsilon^{3/2}$ and the following expression is given for γ by Dukhin et al *(26)* :

$$\gamma = \frac{\varepsilon ab}{4\,\kappa}\ [exp(F\varphi_\delta/2RT)\text{-}1] \qquad (9)$$

where φ_δ is the Stern potential and κ^{-1} the Debye-Hückel length which is proportional to $\varepsilon^{1/2}$.

The acrylamide dipole moment which is 3.97 D in apolar solvents *(27)* is significantly larger than the dipole moment of pure water (1.8 D and 3.11 D in the gaseous and liquid states respectively). Despite the large dipole moment of acrylamide, the dielectric increment of acrylamide water mixtures is negative *(28)* ($d\varepsilon/dC = -0.8\ 1$ mole^{-1}). The dipole moment of liquid water is largely due to the strength of O-H...O hydrogen bonds which make water strongly structured and in the presence of the highly polar amide the ordering of water is disrupted. Thus one expects a severe decrease of the dielectric constant of water near the surface of the colloid where the concentration of polymer is large. Although it is impossible from our data to estimate precisely the polymer layer thickness, an order of magnitude can be conjectured : for $C_0 = 10^{-3}$ g/l, the hydrodynamic thickness estimated from the rotary diffusion coefficient is approximately 80 nm (Figure 6). It has been shown that the hydrodynamic thickness for a θ solvent cannot be less than twice the geometrical thickness *(29)*. Therefore, taking 40 nm as a rough estimate for the average distance of monomers from the surface (40 nm is probably for polyacrylamide in water, a good solvent, a lower bound) and since κ^{-1} is 100 nm, we infer that a large part of the ion atmosphere is "located" within the polymer layer and that the effect on the dielectric properties of the solvent might be strong.

ii) According to Henry, the electrophoretic mobility U_E of a colloid is proportional to the ζ potential at the shear plane and to ε :

$$U_E = \varepsilon \; \zeta \; \eta^{-1} \; f(\kappa R) \tag{10}$$

$f(\kappa R)$ being a correction for the particle dimension R with respect to the ion layer thickness (30) and η the viscosity of the medium. In the presence of a polymer layer, the electrophoretic mobility is best described in terms of a modified Navier-Stokes equation which expresses the velocity profile of the solvent near the particle through a second order differential equation (31-33) :

$$\eta \; \frac{d^2v(x)}{dx^2} - \tau(x)\,f\,v(x) = E\,\rho(x) \tag{11}$$

Equation 11 is Smoluchowsky's equation including an additional friction force due to the polymer. v(x) is the fluid velocity parallel to the particle surface at location x, the first term on the left is the viscous drag, the second is the force imparted to the fluid by the monomer beads, f is a friction coefficient and $\tau(x)$ is the monomer density. The right hand term is the electrical volume force which is proportional to the field E and to the net charge $\rho(x)$ per unit volume. The electrophoretic mobility U_E, which is the limit of v(x)/E for $x >> \kappa^{-1}$, can be expressed in the following form (see Appendix) :

$$U_E = U_E^0 F(bl^{-1}, b\kappa) \tag{12}$$

U_E^0 is the mobility in the limit of vanishing small frictional interactions and this term is proportional to ε. The function F describes frictional effects in terms of the adimensional parameters bl^{-1} and $b\kappa$, b being the polymer layer thickness and l the Debye-Buche shielding length ($l^2 = \eta/f\tau_0$ where τ_0 is the density at x = 0, for more details see ref.(31)). According to equation 12, dielectric change and frictional interactions between monomers decrease the mobility of the particle when the latter acquires a net velocity with respect to the solvent. The function F in equation 12 expresses the mobility reduction of a fluid in a semi-permeable medium. This factor cannot become zero, because complete blocking of the fluid is not expected in a semi-permeable medium (detailed calculations of v(x) have been performed for different concentration profiles (31-32)). However, the mobility U_E is seen in Figures 3 and 5 to become zero at moderate polymer surface concentrations for Mw = 1.2×10^6 . Since such large effects may only be ascribed to friction effects when $\kappa^{-1} << b$, which is not the case here, it is clear that the mobility decrease cannot be solely described in terms of viscous dissipative processes.

iii) Dielectric properties could explain the sharp decrease of α_S and U_E which occurs when C_S becomes larger than 4.10^{-3} g/g (Figure 3). At small C_S, α_S passes however through a distinct maximum (Figures 3 and 4), which indicates that several concomitant mechanisms are operating on the molecular level. Some effects have thus far not been considered. For instance, if surface binding involves the amide group, a dipole layer is build up on the surface in which individual dipoles might adopt a preferential orientation. It is then easy to show that the surface potential φ_0 depends on the number of dipoles in the "train" sequences of monomers near the surface :

$$\varphi_0 = \varepsilon^{-1}\left[n \ \mu + \int_0^\infty x \ \rho(x) \ dx\right] \qquad (13)$$

where n is the number of dipoles of strength μ per unit surface and the last integral is the contribution of mobile ions to φ_0, a quantity proportional to the product $\sigma\kappa^{-1}$ of surface charge σ and electrical layer thickness κ^{-1}. Depending on the sign and magnitude of μ, the absolute value of φ_0 could increase with respect to its value in the absence of polymer. A large surface potential indicates that counterions are spread over a large volume and according to equation 9 the dipole moment then increases, but if this effect plays a role, it is important at low polymer concentrations when most polymer molecules are "lying" flat on the surface.

Hydrodynamic Thickness. In Figure 7 is represented as a function of polymer concentration the hydrodynamic thickness L_r determined for a series of molecular weights by electro-optics through equations 6 and 7. There is a distinct dependence on chain length over the entire concentration range. This is at variance with the electrophoretic mobility behaviour reported in Figure 5. Since hydrodynamic forces are long range forces and the parameter L_r expresses the effect of the largest loops (5,29), the strong dependence on chain length is to be expected. On the other hand, electrophoretic mobility as discussed before allies frictional and electrical properties and the change in the distribution $\rho(x)$ in equation 11 which appears after polymer adsorption could be less polymer weight dependent.

In Figure 8 is reported the dependency of L_t on chain length measured in the plateau region of the adsorption isotherm. The following power law is derived :

$$L_t = 7.63 \ x \ 10^{-2} \ Mw^{\,0.55} \qquad (14)$$

The exponent $\alpha = 0.55$ agrees well with previous results from inelastic light scattering in other colloid-polymer systems : $\alpha = 0.56 -0.54$ for poly(oxyethylene) adsorbed on polystyrene latex and precipitated silica (34-35) ; $\alpha = 0.50$ for poly(vinylalcohol) adsorbed on polystyrene latexes (36) ; $\alpha = 0.55$ for polyacrylamide adsorbed on modified silica (37). We remark that all the exponents are close to the exponent of the hydrodynamic radius R_h of polymers in good solvents ($R_h \simeq M^{-\nu}$ with ν of the order of 0.53-0.51, for more information on this point the reader should consult ref.(38)). This by no means implies that polymers are adsorbed in their solution conformation. The significance of the effective hydrodynamic thickness of a solid particle coated with a polymer layer can be analyzed via equation 11 with the electrical force replaced by a pressure gradient. This equation was first solved for an exponential distribution (29). De Gennes has since derived a solution with a concentration profile obtained from scaling arguments (5) :

$$\tau(x) \simeq x^{-4/3} \qquad (15)$$

A local correlation length $\xi(x)$ was defined (the length over which excluded volume effects are screened), the polymer layer at the interface then resembling a network of variable mesh size $\xi(x)$. When the hydrodynamic equation is integrated up to a distance corresponding to the longest mesh size R_F (the Flory radius of polymers in solution), it is found that the hydrodynamic thickness scales

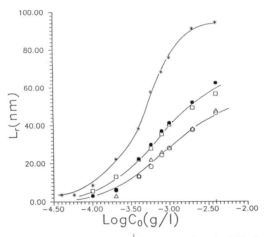

Figure 7.:   Hydrodynamic length L_r from rotational diffusion coefficient
against log of initial concentration C_0 (*) Mw $= 1.2 \times 10^6$;
(●) 5.93×10^5 ; (□) Mw $= 3.6 \times 10^5$; (△) Mw $= 3.17 \times 10^5$;
(O) Mw $= 3.03 \times 10^5$.

like the radius of polymers in solution, i.e., like $M^{3/5}$. The exponent found experimentally is actually a little less than 0.6. Such a discrepancy in the exponent between static and dynamic dimensions has also been found for the coil size in solution, the dynamic exponent of polymers in good solvents being of the order of 0.53-0.55. It was suggested that the origin of the difference lay in the computation of the average inverse bead distance which reaches its asymptotic value only for very long chains, or in the present case very large loops (*39*).

In Figure 9 is reported the logarithmic dependence of L_r on the molecular weight :

$$L_r = 4.28 \times 10^{-2} \, Mw^{\,0.55} \tag{16}$$

The difference in prefactor as compared to relation 14 is due to the fact that the L_r values were recorded at a bulk concentration C_0 of 4.10^{-3} g/l at which adsorption had not reached its saturation value.

Conclusion

Electro-optical measurements provide with respect to classical electrophoretic techniques a complementary means of obtaining useful information on the electrical surface properties of colloids in the presence of a dense and thick surrounding polymer layer. In our study, it is shown that the adsorption of a neutral polymer, polyacrylamide, modifies considerably the structure of the ion atmosphere around the colloid already at very low adsorption. Emphasis is laid on the particular dielectric properties of the acrylamide moieties in interpreting the observed effects and further investigations in other polymeric systems should demonstrate whether these effects are characteristic of this system or not. In the electro-optical method, steady-state and transient observations can be performed with the advantage of separating effects originating solely from the structure of the ion layer and dissipative effects originating from the viscous frictional interactions of the solvent with the monomeric beads. The latter give via determination of the rotary diffusion coefficient information concerning the structure of the polymer layer. In this work, the hydrodynamic thickness of the polymer layer was obtained from the rotary diffusion coefficient for a series of molecular weights and compared to the thickness obtained independently from the translational diffusion coefficient by inelastic light scattering. The power law exponents characterizing the molecular weight dependence were found to be identical by the two techniques. This exponent agrees well with the idea that the surface layer structure can be described as a continuous profile built on the concept of a local correlation length similar to the excluded volume effect for polymers in a good solvent.

Appendix

To integrate equation 11 we expand $v(x)$ in a power series of f :

$$v(x) = v_0(x) + f \, v_1(x) + f^2 \, v_2(x) \, ... \tag{A17}$$

By substitution in equation 11 and making all coefficients of identical power in f vanish, we obtain :

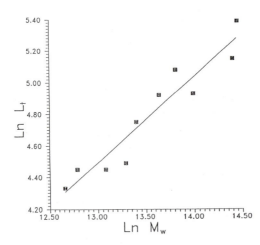

Figure 8.:   Hydrodynamic length L_t as a function of molecular weight.

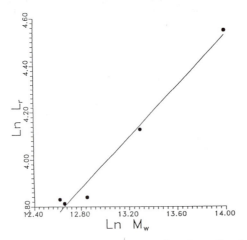

Figure 9.:   Hydrodynamic length L_r as a function of molecular weight.

$$\eta\, v_0^\circ(x) = -E\rho(x)$$
$$\eta\, v_1^\circ(x) = \tau(x)v_0(x)$$
$$\dotsb \tag{A18}$$
$$\eta v_n^\circ(x) = \tau(x)v_{n-1}(x)$$
$$\dotsb$$

In the Debye-Hückel approximation, we have :

$$\rho(x) = \varepsilon\kappa^{-2}\varphi_0\, exp(-\kappa x) \tag{A19}$$
$$v_0(x)/E = U_E^0[1-exp(-\kappa x)] \tag{A20}$$
$$U_E^0 = \varepsilon\varphi_0/\eta \tag{A21}$$

The factor U_E^0 repeats in each equation, and this allows for the form of equation 12. If $\tau(x)$ is known, $v(x)$ can be reached by successive integration. Explicit expressions were derived for an exponential distribution (*31*) :

$$U_E = U_E^0\, A(b\kappa, bl^{-1}/B(bl^{-1})$$

The functions A and B are expressed as infinite series of the adimensional parameters $b\kappa$ and bl^{-1}

$$A = (b\kappa)^2 \sum_{n=0}^{\infty} (bl^{-1})^{2n}/\prod_{i=0}^{n} (n-i+b\kappa)^2$$

$$B = \sum_{n=0}^{\infty} (bl^{-1})^{2n}/(n!)^2$$

De Gennes, using a scaling law for the concentration profile and a self-similar structure of adsorbed chains, has derived the apparent Zeta potential for small and large values of $b\kappa$ (*32*).

Literature Cited

(1) Fredericq, E.; Houssier, C. *Electric Dichroism and Electric Birefringence*, Clarendon Press:Oxford, 1973.
(2) Stoylov, S.P. *Colloid Electro-Optics, Theory, Techniques and Applications*, Colloid Science; Academic Press:London, 1991.
(3) Napper, D.H. *Polymeric Stabilization of Colloid Dispersion*, Colloid Science; Academic Press: New york, 1983.
(4) Vincent, B.; Whittington, S.G.; *Surface and Colloid Science*, Matijevic, E., ed.; Plenum Press: New York, 1982; Vol.12; p.14.
(5) De Gennes, P.G. *Macromolecules* **1982**, *14*, 1637; ibid *15*, 492.
(6) Déjardin, P.; Varoqui, R. *J. Chem. Phys.* **1981**, *75*, 4115.
(7) Zocher, H.; Heller, W. *Zeitschift für Anorg. and Allg. Chemie* **1930**, *186*, 75.
(8) Matijevic, E.; Scheiner, P. *J. Colloid Interface Sci.* **1978**, *63*, 509.
(9) Paterson, R.; Rahman, H. *J. Colloid Interface Sci.* **1983**, *94*, 60.
(10) Parks, G.A.; De Bruyn, P.L. *J. Phys. Chem.* **1962**, *66*, 967.
(11) Onoda, G.Y.; De Bruyn, P.L. *Surface Sci.* **1966**, *4*, 48.

(12) Furstenau, D.W.; Healy, J.W. *Principles of Mineral Flotation*, Academic Press: New York, London, 1972; Chapter 6.
(13) Schultz, R.C.; Cherdron, H.; Kern, W. *Makromol. Chem.* **1958**, *28*, 197.
(14) François, J.; Sarazin, D.; Schwartz, T.; Weill, G. *Polymer* **1979**, *20*, 969.
(15) Pefferkorn, E.; Carroy, A.; Varoqui, R. *J. Polym. Sci., Polym. Phys. Ed.* **1985**, *23*, 1997.
(16) Stoylov, S.P. *Adv. Colloid Interface Sci.* **1971**, *3*, 45.
(17) Petkanchin, I.; Brückner, R.; Sokerov, S.; Radeva, Ts. *Colloid Polym. Sci.* **1979**, *257*, 160.
(18) Perrin, F. *J. Phys. Radium* **1934**, *5*, 497.
(19) Schmitz, K.S. *Dynamic Light Scattering by Macromolecules*, Academic Press: New York, 1990.
(20) Pefferkorn, E.; Jean-Chronberg, A.C.; Chauveteau, G.; Varoqui, R. *J. Colloid Interface Sci.* **1990**, *137*, 66.
(21) Petkanchin, I.; Suong, T.T. *J. Colloid Interface Sci.* **1985**, *108*, 553.
(22) Buleva, M.; Stoimenova, M. *J. Colloid Interface Sci.* **1991**, *141*, 426.
(23) Stoimenova, M.; Radeva, Ts. *J. Colloid Interface Sci.* **1991**, *141*, 433.
(24) Radeva, Ts.; Petkanchin, I.; Varoqui, R.; Stoylov, S. *Colloids Surf.* **1989**, *41*, 353.
(25) Radeva, Ts.; Stoimenova, M. *Colloids Surf.* **1991**, *54*, 235.
(26) Dukhin, S.S.; Shilov, V.N.; Stoylov, S.; Petkanchin, I. *Ann. de l'Univ. Sofia*, **1968/69**, *63*, 125.
(27) Mc Clellan, A.L. *Table of Dipole Moments*, Rahowa Enterprises, 1974; Volume 2.
(28) Cohn, E.J.; Edsall, J.T. *Proteins, Amino-Acids and Peptides*, Rheinhold Publ. Corp: New York, 1950.
(29) Varoqui, R.; Déjardin, Ph. *J. Chem. Phys.* **1977**, *66*, 4395.
(30) Hunter, R.J. *Zeta Potential in Colloidal Science*, Academic Press: London, New York, 1981.
(31) Varoqui, R. *Nouveau Journal de Chimie* **1982**, *4*, 187.
     Equation 6 in that paper should be corrected for a misprint in the sign of the denominator, cf to the discussion of the Appendix here.
(32) De Gennes, P.G. *C.R. Acad. Sci. Paris* **1984**, *299*, serie II, *14*, 913.
(33) Cohen Stuart, M.A.; Waajren, F.W.W.H.; Dukhin, S.S. *Colloid Polym. Sci.* **1984**, *262*, 423.
(34) Killmann, E.; Maier, H.; Baker, J.-A. *Colloids Surf.* **1988**, *31*, 51.
(35) Kato, T.; Nakamura, K.; Kawaguchi, M.; Takahashi, A. *Polymer J.* **1981**,
(36) Garvey, M.J.; Tadros, Th.F.; Vincent, B. *J. Colloid Interface Sci.* **1974**, *49*, 57.
(37) Varoqui, R.; Pefferkorn, E. *Prog. Colloid Polym. Sci.* **1990**, *83*, 96.
(38) De Gennes, P.G. *Scaling Concepts in Polymer Physics*, Cornell University Press; Ithaca, London, 1979; Chapter 6.
(39) Weill, G.; des Cloiseaux, J. *J. Phys.* (Les Ulis, Fr) **1979**, *40*, 99.

RECEIVED May 10, 1993

FLOCCULATION AND STABILIZATION

Chapter 12

Interaction between Polyelectrolytes and Latices with Covalently Bound Ionic Groups

H. Dautzenberg, J. Hartmann, G. Rother, K. Tauer, and K.-H. Goebel

Max-Planck-Institute of Colloid and Interface Research, O–1530
Teltow-Seehof, Germany

The interaction between latex particles with covalently bound ionic groups and polyelectrolytes was studied by dynamic and static light scattering. In the systems investigated two effects occurred: a coating of the latex particles with a closed-fitting polyelectrolyte layer and coagulation, caused by interparticle bridging. The aggregation effects could be suppressed by a large excess of the polycation in highly diluted systems. The amount of bound polyelectrolyte was determined via a subsequent polyelectrolyte complex formation. Possible determinants of the structure of the polyelectrolyte layer are discussed.

It is a well-known fact that in highly diluted systems latex particles may be covered with a layer of water-soluble polymers, preventing coagulation. This makes it possible to study the structural parameters of the polymer layer by static and dynamic light scattering. Killmann et al.(1) could show that the thickness of the adsorbed layer of poly(ethyleneoxide)s on polystyrene latexes depends strongly on the amount of the adsorbed polymer and still more on the molecular weight of the polymer. The strong Coulombic interaction between a charged latex surface and an oppositely charged polyelectrolyte impedes the suppression of coagulation effects. Main objectives of this paper are :
 - coating of a charged latex with a polyelectrolyte without coagulation
 - study of the structural parameters of the polymer layer , using static and dynamic light scattering
 - determination of the amount of bound polyelectrolyte by subsequent polyelectrolyte complex formation

0097–6156/93/0532–0138$06.00/0

- modification of the structure of the polymer layer by variation of the charge density of the latex.

Materials

To exclude effects of desorption of the ionic groups on the latex surface during the interaction with an oppositely charged polyelectrolyte we used a latex with covalently bound ionic groups.

The latex was synthesized by emulsion polymerization employing a non-ionic initiator and a monomer emulsifier with sulfonate groups (*2,3*), which is primarily incorporated at high degrees of conversion (Figure 1), leading to high charge density on the surface of the latex particles. Conditions of polymerization see Table 1. The latex sample was characterized by turbidimetry, light scattering, electron microscopy and conductometric titration.

The polycation used to cover the latex was poly(diallyl-dimethyl-ammonium chloride). As polyanion in PEC formation Na-poly(styrene sulfonate) was employed. To vary the charge density by blocking of the anionic charge of the latex we applied the cationic surfactant Hyamine 1622. The characteristics of all samples used in this work are given in Table 2.

Methods of investigation

Static and dynamic light scattering as well as electron microscopy were used for structural investigations. By a combination of these methods one may determine the mass, the radius of gyration, the hydrodynamic radius and the polydispersity of a particle system. By a comparison of the results obtained for the original and the polymer covered latex the structural parameters of the polymer layer can be assessed. Ultracentrifugation was used for latex separation. A subsequent determination of the polycation concentration via PEC-formation in the supernatant yields exact information on the amount of polycation, which was bound on the latex.

Experimental

All experiments were carried out by mixing of the latex and the polymer solutions directly in the scattering cell of the light scattering instrument, a Sofica 42 000 (Wippler and Scheibling, Strasbourg, France) under gentle stirring, with continuous registration of the light scattering intensity at fixed angle ($\Theta = 45^\circ$). As solvent saltfree water, purified by repeated filtration

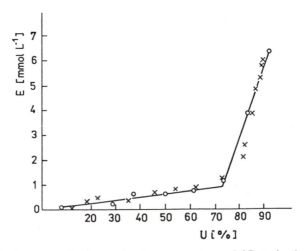

Figure 1. Amount of polymerized monomer emulsifiers in dependence on the degree of conversion.
Styrene with Na-sulfopropyl-octadecyl-maleate (○)
               Na-sulfopropyl-tetradecyl-maleate (x).

Table 1: Condition of synthesis of latex:

Emulsion polymerization:
$\sim$ 10 h at t = 60^0 C under stirring

Composition:

water:	250 g
styrene:	50 g
Na-sulfopropyle-tetradecyl-maleate:	6.85 g
(monomer emulsifier), M = 456.6	
Azoisobutyronitrile:	436 mg
(non-ionic initiator)	

and ion exchange, was used. The solutions were made dust-free by filtration through 0.2 μm-membrane filters. Addition of solutions into the scattering cell was carried out via an immersed capillary with very slow dosage rates. Periodic interruptions of the dosage were made so that the complete angular dependence of the scattering intensity could be obtained in various intermediate states. Dynamic light scattering experiments were carried out with the instrument SIMULTAN (ALV, Langen, Germany). The correlation functions were analyzed by a third order cumulant fit. Latex separation was performed by 4 hour centrifugation at 50 000 revol./min with a preparative ultracentrifuge (Janetzki, Germany).

Results

Latex coating.   The first objective was to cover the latex with the polycations without coagulation. Cationic surfactants as well as polycations are normally used as titrants for the determination of the charge density of anionically stabilized latexes (*4*). In an appropriate range of concentration, the point of charge compensation is indicated by clearly pronounced flocculation.

In a first experiment we added to 10 mL of a latex solution of concentration $c_L = 8.15 \ 10^{-5}$ g cm^{-3} the polycation (PC) solution of $c_C = 1 \ 10^{-5}$ g cm^{-3} with a dosage rate $v_D = 2$ mL/70 min. Fig. 2 shows the dependence of the reduced scattering intensity I_{45}/c_L (normalized to 1 for the pure latex) on the added volume of the PC-solution. After the addition of 1.8 mL PC-solution, flocculation occurred. This corresponds exactly to a 1:1 charge stoichiometry in the system, because the number of charges N_e of the latex and the polycation respectively is given by $N_e = N_A c V m_e^{-1}$, where N_A - Avogadro's number , c - concentration, V - volume, m_e - molar mass per charge unit (see Table 2). The increase of the I_{45}/c_L-values indicates a coagulation of the latex already at small contents of polycations. The scattering curves, registered after stepwise addition of the polycation solution are given in Figure 3 as the Guinier plot ln (R(h)/(K c)) vs. sin$^2(\Theta/2)$, where R(h) - Rayleigh ratio of the scattering intensity, K - contrast factor, c - mass concentration of the scattering species, h = $4\pi/\lambda$ sin($\Theta/2$), λ - wave length in the medium, Θ -scattering angle. The increasing level and curvature of the scattering curves confirm the increasing coagulation with addition of polycation.

The addition of the latex into the PC solution should provide a better means of covering the latex completely with polycation without coagulation. Therefore, we added to 10 mL PC solution of concentration $c_C = 2 \ 10^{-6}$ g cm^{-3} a latex solution of $c_L = 9.35 \ 10^{-5}$ g cm^{-3} with a dosage rate $v_D =$

Table 2: Materials

Latex:

    Polystyrene with covalently bound Na-sulfonate groups

    Radius: a $= 40 \pm 2$ nm

    Polydispersity: relative standard deviation $\sigma = 0.1$

    Surface charge density: $\rho = 18$ μC cm^{-2} $= 1.12$ 10^{14} e cm^{-2}

    Molar mass per charge unit: $m_e^L = 7350$

    Refractive index increment: dn/dc $= 0.243$ cm^3 g^{-1}

Polycation:

    Poly(diallyl-dimethyl-ammonium chloride)   (PDADMAC)

    $M_w = 1.4$ 10^5 g mol^{-1} , $M_w/M_n = 2$

    $m_e^C = 161.5$,   dn/dc $= 0.193$ cm^3 g^{-1}

Polyanion:

    Na-Poly(styrene sulfonate)   (NaPSS)

    $M_w = 2.0$ 10^5 g mol^{-1} , $M_w/M_n = 1.1$

    $m_e^A = 204$,   dn/dc $= 0.192$ cm^3 g^{-1}

Cationic surfactant:

    Hyamine 1622, Switzerland (C_{27} H_{42} Cl NO_2 . H_2O)

    $m_e^S = 466$,   dn/dc $= 0.192$ cm^3 g^{-1}

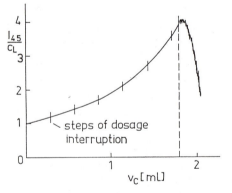

Figure 2. I_{45}/c_L of the system latex/polycation (normalized to 1 for the pure latex) in dependence on the volume of the added PC-solution.

4 mL/70 min. However, the results obtained for the reduced scattering intensity I_{45}/c_L (Figure 4) again show coagulation and flocculation, also indicated by the curvature of the scattering curves (Figure 5) at intermediate states. Macroscopic flocculation occurred before charge compensation was reached (theoretically at $V_L = 9.7$ mL). On the other hand, the extrapolation of the curve in Figure 4 to $V_L \Rightarrow 0$ shows that I_{45}/c_L tends to the value of the non-coagulated latex. Consequently, it may be expected that at very low concentration of the latex and very high excess of polycations coagulation can be suppressed. To check this, we investigated several systems , obtained by slow addition of highly diluted latex to PC solutions of differently high excess concentrations of polycation. It must be notized here that "excess of polycations" means an excess of the cationic charges in the system and not of the mass concentration. Table 3 gives the results of static and dynamic light scattering measurements on such systems. With increasing PC concentration a decrease of the radius of gyration and the hydrodynamic radius was observed, while the M_w-value remained nearly constant, i.e. in all systems coagulation plays a negligible role. The increase of the mass of the latex particles by covering with polycations corresponding to a 1:1 stoichiometry is in the range of two percent (see the different m_e-values), which is in the limit of the accuracy of preparation and measurement. At very high excess of polycations the radius of gyration and the hydrodynamic radius approach to the values of the pure latex. This result can only be explained by a coating of the latex with a compact layer of polycations.

The scattering curves of the systems 1, 2 and 6 of Table 3 show that coagulation is completely excluded only in the system with the highest PC concentration (Figure 6a). The angular dependence of the apparent diffusion coefficients reflects the same result (Figure 6b), but the sensitivity of the shape of the static scattering curve to coagulation is higher.

In principle also the ratio of the second cumulant to the square of the first one (μ_2/Γ_1^2 - relative quadratic standard deviation of the size distribution) should give an information about changes of the polydispersity due to coagulation. However, at the scattering angle 90°, μ_2/Γ_1^2 lies in the range of 0.03 ± 0.02 and no systematic alterations could be detected. In the small angle range the fluctuations of the μ_2/Γ_1^2 - values are too large to draw reliable conclusions about coagulation.

Determination of the bound amount of polycation. Since the amount of bound polycations could not be calculated from the increase of the mass of the latex particles, we tried to determine it from the concentration change of free polycations in the original solution and after the mixing with latex. PEC formation offers a very sensitive method for a quantitative analysis

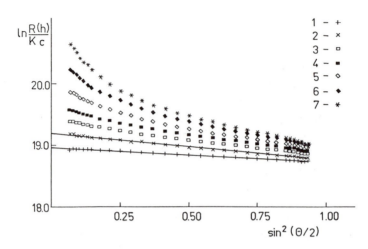

Figure 3. Guinier plot of the scattering curves at different stages of PC-solution dosage (compare Figure 2).

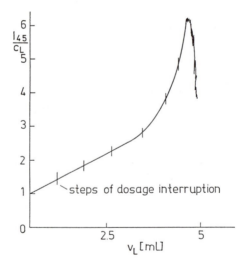

Figure 4. I_{45}/c_L of the system polycation/latex in dependence on the volume of the added latex.

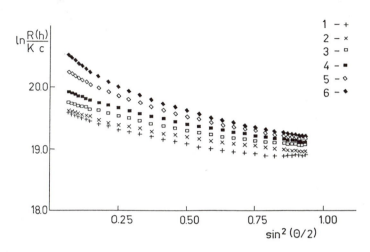

Figure 5. Guinier plot of the scattering curves at different stages of latex addition (compare Figure 4).

Table 3: Results of light scattering measurements on coated latex at high excess of cationic charges. Addition of 8 mL latex ($c_L = 1\ 10^{-5}$ g cm^{-3}) to 8 mL PC-solutions of different concentrations (v_D = 4 mL/70 min), $a_z = \ <a^2>_z^{1/2} = (5/3\ <s^2>_z)^{1/2}$

$c_C\ 10^6$ [g cm^{-3}]	$M_w\ 10^{-8}$ [g mol^{-1}]	a_z [nm]	R_H [nm]
0		43.1	42.2
1		56.0	46.7
2.5	1.7 ± 0.1	52.0	43.8
5		48.7	43.5
10		46.4	42.1
25		43.0	42.2

(5,6). During the mixing of strong oppositely charged polyelectrolytes the growth of PEC particles started immediately with the mixing process. The structural parameters of the PEC particles do not change significantly with rising degree of conversion (X) up to about X = 0.9. In many cases flocculation occurs at the 1:1 charge stoichiometry of the polyelectrolyte components. This is demonstrated by the scattering curves of PEC particles between PDADMAC and NaPSS in highly diluted systems (c_C° = 2.5 10^{-5} g cm^{-3}, c_A° = 5 10^{-5} g cm^{-3}) at different degrees of conversion (Figure 7). While the scattering curves up to X = 0.9 are quite similar, a sharp transition to a highly aggregated system occurs at full conversion, which is indicated by a change of the scattering intensity of about a factor 50 and a strong curvature of the scattering curve.

To determine the fraction of polycations bound by the latex, we added to 8mL PC solution of a concentration c_C = 5 10^{-5} g cm^{-3}, 8 mL latex with concentrations of c_L = 5 10^{-4} g cm^{-3} and 1 10^{-3} g cm^{-3} respectively, corresponding to 22% and 44% of the charges of polycations. After separation of the coated latex by ultracentrifugation, the PC concentration was determined by PEC formation with NaPSS (c_A = 5 10^{-5} g cm^{-3}) in comparison with a pure PC solution. The titration curves are given in Figure 8. Curve 1 corresponds to the titration of 8mL pure PC solution of a concentration c_C = 2.5 10^{-5} g cm^{-3}. The full lines mark the points of 1:1 stoichiometry, taking into account the binding of polycations on the separated latex. Flocculation occurs exactly at these points, indicating a covering of the latex with a PC layer according to the anionic groups on the latex surface. The reliability of the results obtained may be assessed by a comparison with the flocculation point, which should be expected for the supernatants without binding of the polycations on the latex particles (dashed line). This line is somewhat shifted in regard to the pure PC solution, because only 7.5 mL of the supernatants were titrated.

PEC formation was also carried out in the presence of coated latex. Three systems were prepared by mixing 10mL PC solution (c_C = 1 10^{-5} g cm^{-3}) with 0.5, 1.0 and 2.0 mL latex (c_L = 1 10^{-3} g cm^{-3}) respectively, corresponding to 11, 22 and 44% binding of the polycation. 8 mL of the mixed systems were then titrated with NaPSS solution (c_A = 2.5 10^{-5} g/cm^{-3}, v_D = 4 mL/70 min). Even in this case the coagulation points are clearly pronounced (Figure 9), although the scattering level of the latex particles is much higher. The dashed lines in Figure 9 mark the flocculation points corresponding to the total amount of polycation in the systems and the full lines correspond to the flocculation for the amount of free PC. The experimentally observed flocculation points are in very good agreement with the amount of free PC, suggesting that the polycations bound on the latex do not take part in PEC formation.

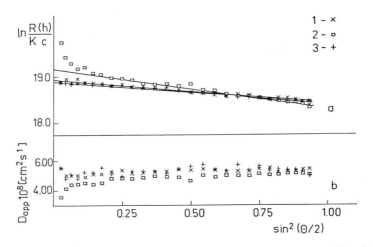

Figure 6. Scattering curves (a) and diffusion coefficients (b) of the systems 1 (1), 2 (2) and 6 (3) of Table 3.

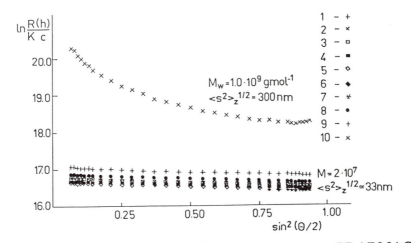

Figure 7. Scattering curves of PEC particles between PDADMAC and NaPSS at different degrees of conversion X = 0.1 (0.1) 1.

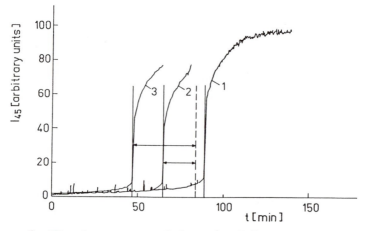

Figure 8. Titration curves of the pure PC solution - (1), the supernatants after latex separation: $c_L = 5 \ 10^{-4} \ g \ cm^{-3}$ - (2), $c_L = 1 \ 10^{-3} \ g \ cm^{-3}$ - (3) with NaPSS ($v_D = 4 \ mL/70 \ min$).

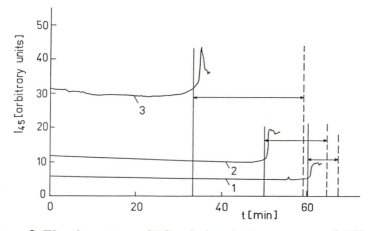

Figure 9. Titration curves of PC-solutions in the presence of different amount of latex: 1 - 0.5 mL, 2 - 1.0 mL, 3 - 2.0 mL.

One can conclude that the polycations are bound on the latex surface according to charge compensation. The binding is so strong that a close-fitting layer results and the bound polycations are inactive in subsequent PEC formation. Electron microscopic pictures of the original and the coated latex (Figure 10) show only a more rough surface of the polymer covered particles. These results appear reasonably due to the comparable average charge distances on the latex surface of about 1 nm and along the PC chains of 0.5 nm and the remaining mobility of the anionic charge centres, due to the spacing of the sulfonate groups by the propyle groups.

Variation of the charge density on the latex surface. The last objective of this paper is the variation of the structure of the polycation layer, what may be achieved in the following ways:
(1) - reduction of the charge density on the latex surface,
(2) - change of the conformation of the free polycation by increasing the ionic strength of the medium (addition of salt),
(3) - variation of polycation molecular weight, charge density and structure.
Only the results of variation of the latex surface charge density are presented here.
Reduction of the ionic monomer emulsifier during the synthesis of the latex led to latex coagulation. Therefore, we tried to block a part of the anionic groups with the cationic surfactant Hyamine. This surfactant is used as a titrant for the determination of the charge density of anionically stabilized latexes, i.e. it causes coagulation. However, the coagulation behavior depends strongly on the concentration of the component solutions and on the rate of mixing. It could be shown that in highly diluted systems, and with slow addition of the Hyamine solution, the latex can be covered completely by the surfactant without coagulation effects. The reduced scattering intensity of a latex solution (V = 10 mL, c_L = 0.935 10^{-4} g cm^{-3}) during mixing with Hyamine solution (c_H = 2.5 10^{-5} g cm^{-3}, v_D = 2 mL/70 min) is given in Figure 11. Up to 1:1 charge stoichiometry in the system only a slight increase of the reduced scattering intensity occurs, which corresponds primarily to the increase of the particle mass from covering with surfactant. The system remains stable beyond the charge compensation point. Figure 12 confirms this result, revealing no changes of the shape of the scattering curve in the whole range investigated.

Therefore, charge blocking of the latex by low molecular cations makes it possible to vary the surface charge of the latex to a great extent without coagulation. However, the interaction of such partially neutralized latex particles with the polycation led again to compact polyelectrolyte layers.

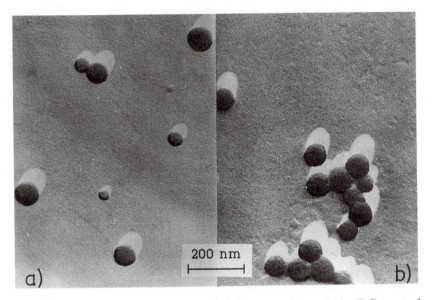

Figure 10. Electron micrographs of the pure (a) and the PC coated latex (b).

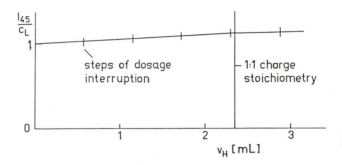

Figure 11. Light scattering intensity I_{45}/c_L vs. the added volume during the coating with Hyamine.

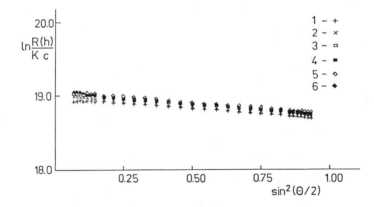

Figure 12. Guinier plot of the scattering curves at different stages of Hyamine dosage (compare Figure 11).

This result may suggest a displacement of the Hyamine from the latex surface by the polycation, but up to now we could not check it explicitly

Conclusions

Latex particles with covalently bound ionic groups can be used to study the interaction between charged latexes and polyelectrolytes without charge desorption effects. Static and dynamic light scattering experiments showed that the interaction leads to a coating of the latex particles with polyelectrolyte, but also to an interparticular bridging and coagulation. The aggregation phenomena can be suppressed by slow addition of highly diluted dispersions to polyelectrolyte solutions with high   excess of polyelectrolyte. The interaction between highly charged particles and a strong polyelectrolyte results in a close-fitting polymer layer. The amount of the bound polyelectrolyte corresponds to a 1:1 charge stoichiometry and the bound polyelectrolyte is inactive in subsequent polyelectrolyte complex formation. The surface charge density of the latex may be greatly varied by blocking with a low molecular, oppositely charged surfactant.

Acknowledgements

We thank Dr. Tesche , Fritz-Haber-Institute of the MPG of Germany, Berlin-Dahlem, for the electron microscopic studies.

References

1. Killmann, E., Sapuntzjis, P., Maier, H.; <u>Makromol. Chem., Macromol. Symp.</u>, 1992, <u>61</u>, 42
2. Tauer, K., Goebel, K.H., Kosmella, S., Neelsen, J. and Staehler, K.; <u>Plaste und Kautschuk</u>, 1988, <u>35</u>, 373
3. Tauer, K., Goebel, K.H., Kosmella, S., Staehler, K. and Neelsen, J.; <u>Macromol. Chem., Macromol. Symp.</u>, 1990, <u>31</u>, 207
4. Fischer, J.P. and Noelken, E.; <u>Progr. Coll. Polym. Sci.</u>, 1988, <u>77</u>, 180
5. Philipp, B., Dautzenberg, H., Linow, K.J., Koetz, J. and Dawydoff, W.; <u>Prog. Polym.Sci.</u>, 14 (1989) 91
6. Philipp, B., Koetz, J., Dautzenberg, H., Dawydoff, W. and Linow, K.J.; Applied Polymer Analysis and Characterization, edit. by J. Mitchell, Vol. II, Hanser Verlag 1991, pages 282 - 310

RECEIVED April 13, 1993

Chapter 13

Flocculation of Poly(N-isopropylacrylamide) Latices in the Presence of Nonadsorbing Polymer

M. J. Snowden and B. Vincent

School of Chemistry, University of Bristol, Bristol BS8 1TS, United Kingdom

Crosslinked latex particles were prepared by the free radical polymerisation of N-isopropyl acrylamide (NIPAM) using N/N methylene bisacrylamide as a crosslinking agent in water at 70°C. At 25°C the latex had a diameter of 460nm, whilst on heating to 40°C the particles shrunk reversibly to become hard spheres having a diameter approximately 230nm. In the presence of electrolytes at 40°C aggregation of the latex occurred as a result of screening the charges on the particle surface. The same latex at 40°C in the presence of non adsorbing poly(styrene sulphonate) and sodium carboxymethylcellulose flocculated as a consequence of depletion forces operative in the system. The latex redispersed on cooling to 25°C.

A preliminary report (Snowden, M.J. and Vincent, B. J.C.S.Chem Comm, in press) describing the flocculation of poly(N-isopropylacrylamide) latices by the addition of simple electrolyte and non adsorbing poly(styrene sulphonate) illustrated the reversibility of the flocculation of microgel particles under the influence of temperature. It is the objective of this paper to expand on these initial results and to demonstrate the effect of different electrolytes and polymers on particle stability and also to indicate the effect of the polydispersity of the polymer samples on the dispersions.

The critical aggregation temperature of particles of poly(NIPAM) copolymerised with acrylamide was initially studied by Pelton and Chibante (1) using $CaCl_2$. Measurements were made as a function of the fraction of acrylamide incorporated in the polymerisation process and it was found that the temperature at which aggregation was first observed incresed linearly with increasing acrylamide concentration. The critical aggregation temperature of a homopolymer sample of poly(NIPAM) as a function of electrolyte concentration is discussed in this paper along with the valency of the metal cation and

0097–6156/93/0532–0153$06.00/0

the reversibility of the process with respect to temperature is illustrated. We further report the stability behaviour of poly(NIPAM) in the presence of non adsorbing polyelectrolytes in an attempt to illustrate the similarities and differences in the stability of microgel latices compared to the more well known hard sphere colloids and thus illustrate some unique characteristics of poly (NIPAM) latex. Further the effect of polydispersity on the depletion force exerted by non adsorbing polymers is considered, for in a previous publication, Snowden et al (2) studing the effect of non adsorbing poly(styrene sulphonate) and carboxymethyl cellulose on a silica sol noted that for polydisperse samples the crtical flocculation concentration of the polymer became independent of molecular mass and linear charge density. Under such conditions the theoretical models (3,4) developed to describe depletion flocculation were not surprisingly found to be invalid.

Experimental

Poly(N-isopropyl acrylamide) (NIPAM) particles were prepared by the free radical polymerisation of NIPAM in water at 70°C, in the presence of N/N methylene bisacrylamide, according to the procedure described by Pelton and Chibante (1). Following dialysis against distiled water, transmision electron micrographas showed the particles to be monodisperse spheres having a mean diameter of 460nm $\pm$ 12nm. Poly(styrene sulphonate) Mw 46 000 (Mw/Mn 1.05) was purchased from Polymer Laboratories Ltd, Shropshire and a polydisperse sample of poly(styrene sulphonate) having a nominal molecular mass of 50 000 was kindly donated by Unilever. Carboxymethyl cellulose Mw 250 000 and polyethylene oxide Mw 3 000 000, both samples having unquoted polydispersities, were purchased from BDH chemicals Ltd.

The temperature dependency of the particle diameter of poly(NIPAM) latex was determined by photon correlation spectroscopy using a Malvern instruments type 7027 dual LOGLIN digital correlator equipped with a krypton-ion laser ($\lambda = 530.9$nm). The intensity of scattered light was measured at 90°. Particle size was measured from 25°C to 50°C in increments of 5°C both on heating and cooling using an external water bath.

Stability experiments were carried out using dispersions containing 0.1% microgel particles and varying concentrations of sodium chloride, magnesium chloride and lanthinum chloride made up to a total volume of 5 cm^3. The samples were placed in a water bath at 25°C and allowed to stand overnight. The extent of any aggregation in the dispersion was estimated from the wavelength dependence of the turbidity of the dispersions (5), where n = -[d log turbidity/d log wavelength] The same experiment was repeated at 40°C. The stability of similar dispersions of poly(NIPAM) were studied at both temperatures using two different samples of poly(styrene sulphonate), carboxymethyl cellulose and polyethylene oxide as a function of polymer concentration

Results

Figure 1 illustrates the decrease in particle diameter of the poly(NIPAM) latex particles on heating. The initial diameter of 460nm at 25°C decreased to 230nm at 50°C. On cooling however, the particles expanded and adopted their original particle size. This procedure was found to be fully reversible over a number of heating and cooling cycles with no hysterises taking place between the heating and cooling curves. The particles remained dispersed over the temperature range, 25-50°C in water. In the presence of 0.1 mol dm^{-3} NaCl at 25°C the dispersion also remained stable, however on repeating the same experiment at 40°C, aggregation of the poly(NIPAM) latex takes place. The sample flocculated at 40°C at NaCl concentrations in excess of 0.06 mol dm^{-3}, redispersed on cooling to 25°C, having a value for n of -2.6 which is consistent with the particles becoming fully redispersed. This process of heating the samples, inducing flocculation, cooling and gently inverting the dispersion to allow the particles to redisperse can be repeated many times with the process remaining fully reversible on each occasion. Figure 2 illustrates the change in absorbance of the dispersion at 592nm with increasing temperature at several different NaCl concentrations. It can clearly be seen that the onset of aggregation takes place at lower temperatures on increasing the electrolyte concentration in the dispersion. On cooling to 25°C the particles once again slowly redispersed. The critical aggregation concentration of electrolyte at which a 0.1% dispersion of poly(NIPAM) at 40°C destabilised also decreased with an increase in the valency of the cation present. The onset of aggregation took place in 0.06 mol dm^{-3} in NaCl, 0.045 mol dm^{-3} in MgCl$_2$ and 0.02 mol dm^{-3} in LaCl$_3$.

The stability of the poly(NIPAM) particles in the presence of the monodisperse sample of poly(styrene sulphonate) is illustrated in Figure 3. At 25°C the particles remained dispersed up to the highest polymer concentration studied at 0.8%. On repeating the same experiment with the dispersions at 40°C, flocculation was found to take place at polymer concentrations of 0.6% and above. The samples remained flocculated while the temperature was maintained at 40°C. On cooling, the dispersions to 25°C and gently inverting the container several times, the flocculated samples redispersed fully with their "n" values simultaneously decreasing from -1.3 to -2.7. This process was also found to be reversible over a number of heating and cooling cycles. Figure 3 also illustrates the

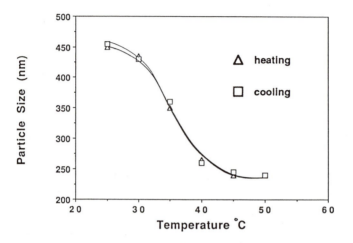

Figure 1.   A plot of particle size against temperature for a
0.01% dispersion of poly(NIPAM) latex in water.

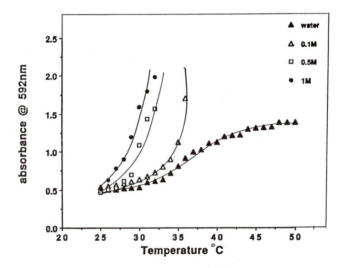

Figure 2.   A plot of turbidity at 592 nm against temperature
for 0.1% dipersion of poly(NIPAM) in different solutions of
NaCl.

same experiment carried out in a background electrolyte of 0.01 mol dm^{-3} NaCl, an insufficient electrolyte concentration to destabilise the particles, the critical flocculation concentration (CFC) of the sodium poly(styrene sulphonate) sample is reduced to 0.35%. The effect of a polydisperse sample of poly(styrene sulphonate) on the stability of a poly(NIPAM) dispersion in water is shown in figure 4. It can clearly be seen that the polydisperse sample induces flocculation at a polymer concentration of.0.75%, a value slightly higher than for the monodisperse sample.

The addition of carboxymethyl cellulose to a poly(NIPAM) dispersion at 40°C results in flocculation of the particles taking place at a polymer concentration of 0.55% in water at pH7 (figure 5). In 0.01 mol dm^{-3} NaCl the CFC is reduced to 0.30% which is consistent with the result obtained for sodium poly(styrene sulphonate) under the same conditions. The dispersions containing carboxymethyl cellulose also redispersed on cooling to 25°C.

The addition of polyethylene oxide (PEO) to the poly(NIPAM) dispersion in a background electrolyte of 0.01mol dm^{-3} NaCl at 40°C did not result in destabilisation of the latex dispersion up to a polymer concentrations of 3.0%.

Discusssion

The decrease in diameter.of the poly(NIPAM) latices on heating from 25°C to 40°C is a consequence of the increase in the χ parameter for N-isopropylacrylamide/water system. This facilitates more polymer-polymer contacts, hence the particles contract and shrink substantially. At 25°C the particles can be envisaged as spongy microgels, swollen with solvent and having very low Hamaker constants. It is not surprising therefore, that no aggregation of the dispersion takes place under the influence of added electrolyte as the van der Waals attraction energy would be very small. At 40°C the particles shrink forcing out any solvent from the interstitial spaces and become very similar to hard spheres in their nature. Aggregation of the particles now takes place as the Hamaker constant increases, and screening the charges at the particle surface allows close enough approach for coagulation to occur.

In the presence of poly(styrene sulphonate) at 40°C the particles flocculate as a consequence of depletion forces operative in the system. It is believed that the polymer does not adsorb onto the particles from water at neutral pH as both species have negative charges. Other hard sphere colloids have been reported to undergo similar depletion flocculation processes with poly(styrene sulphonate) (2,6) under similar conditions but at a fixed temperature. The flocculated latex on cooling down to 25°C redisperses, as a consequence of the particles swelling with solvent and then undergoing conformational rearrangments. The soft nature of the particles at 25°C

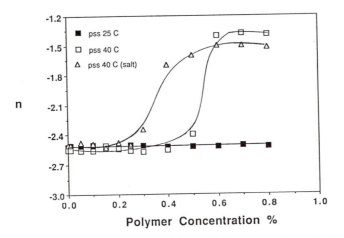

Figure 3.   A plot of d log absorbance/d log wavelength for a 0.1% dispersion of poly(NIPAM) at increasing concentrations of poly(styrene sulphonate) at 25 and 40°C in water at pH7 and in 0.01mol dm^{-3} NaCl.

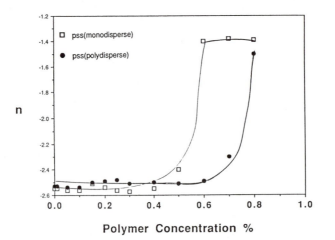

Figure 4.   A plot of d log absorbance/d log wavelength for a 0.1% dispersion of poly(NIPAM) at increasing concentrations of monodisperse and polydisperse poly(styrene sulphonate).

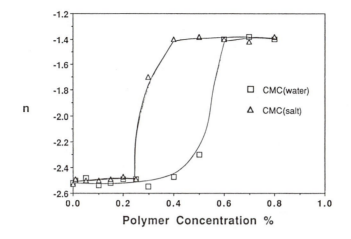

n

Polymer Concentration %

Figure 5.   A plot of d log absorbance/d log wavelength for a 0.1% dispersion of poly(NIPAM) at increasing concentrations of carboxymethyl cellulose at 25 and 40°C in water at pH7 and in 0.01mol dm^{-3} NaCl.

would require very high polymer concentrations to cause destabilisation. Clarke and Vincent (7) investigating the effect of non-adsorbing polystyrene on polystyrene microgels reported the critical flocculation concentration (CFC) of polymer increased significantly as the microgel became swollen and softer. Further in a recent publication Milling et al (8) studying the effect of grafted chain coverage on the depletion flocculation of sterically stabilised silica, reported that the CFC of the free polymer increased with increasing coverage of the grafted polymer layer i.e. "softness" of the particle. In the presence of 0.01 mol dm^{-3} NaCl the electrostatic repulsive potential is greatly reduced as a result of charge screening and therefore flocculation into a secondary minimum resulting from depletion forces takes place more readily. This behaviour may be predicted quantitatively by current theories of colloid stability describing interacting spherical particles (4,9,10). The critical flocculation concentration of the polydisperse polystyrene sulphonate is sligtly higher than a comparable molecular mass monodisperse sample which may be as a result of any shorter polymer chains been able to occupy the interstitial region where polymer chains are normally excluded thus reducing the depletion layer thickness.

As carboxymethyl cellulose is also unlikely to adsorb onto the particle surface due to its negative charge, flocculation is also believed to be as a result of depletion forces. Other negatively charged particles have been shown to undergo depletion flocculation in water with carboxymethyl cellulose at similar concentrations (2). The decrease in the CFC of carboxymethyl cellulose in the electrolyte solution is consistent with that observed with poly(styrene sulphonate) and was not unexpected.

The poly(NIPAM) dispersions containing polyethylene oxide solutions showed no instability at any polymer concentration. Despite carrying out the experiments in 0.01 mol dm^{-3} NaCl where the electrical double layer surrounding the particles, $1/\kappa$ is less than 10nm, thus allowing adsorbed polymer chains to bridge, no instability was observed. Further at higher polymer concentrations there was no evidence of any depletion flocculation operative in the system.

Acknowledgements

M.J.Snowden would like to acknowledge BP Research, Sunbury for providing a research fellowship and to Drs J.Morgan and S.E.Taylor for useful discussions.

Literature Cited

1. Pelton, R.H. and Chibante, P. Colloids and Surfaces, **1986**, 20, 247.

2. Snowden, M.J.; Clegg, S.M.; Williams, P.A.; and Robb, I.D. J. Chem Soc. Faraday Trans, **1991**, 87, 2201.

3. Fleer, G.J.; Scheutjens, J.H.M.H. and Vincent, B. ACS Symp. Ser, **1984**, 240, 245.

4. Vincent, B.; Edwards, J.; Emmett, S. and Croot, R. Colloids and Surfaces, **1986**, 18, 261.

5. Long, J.A.; Osmond, D.W.J. and Vincent, B. J.Colloid Interface Sci, **1973**, 42, 545.

6. Rawson, S.; Ryan, K. and Vincent, B. Colloids and Surfaces, **1989**, 34, 89.

7. Clarke, J. and Vincent, B. J.Chem Soc., Faraday Trans.1. **1981**, 77, 1831.

8. Milling, A.; Vincent, B.; Emmett S. and Jones A., Colloids and Surfaces, **1991**, 51, 185.

9. Derjaguin, B.V. and Landau, L. Acta Physiocochim. U.R.S.S., **1943**, 14, 633.

10. Verwey, E.J.W. and Overbeek, J.Th.G. Theory of the stability of Lyophobic Colloids, Elsiever, Amsterdam, **1948**.

RECEIVED February 23, 1993

Chapter 14

Flocculation of Kaolin Suspensions by Polyelectrolytes

T. M. Herrington, B. R. Midmore, and J. C. Watts

Department of Chemistry, University of Reading, RG6 2AD, United Kingdom

The stability and flocculation of dispersions has been investigated using viscoelastic measurements and dynamic light scattering techniques. The systems studied included kaolinite, titania and polystyrene microspheres. Controlled flocculation of the system was achieved by changing the ionic strength and by adding polyelectrolytes. The viscoelastic behaviour of the suspensions with and without addition of polymer was investigated. Viscometric and oscillatory techniques proved to be powerful methods for studying the binding forces within aggregates and flocs. Also a series of aggregating suspensions was monitored by dynamic light scattering and the rate constants for the coagulation kinetics obtained. The structure factor of the flocs was calculated by computer simulation methods, assuming cluster-cluster aggregation and ballistic particle-cluster aggregation schemes. The results were interpreted in terms of the hydrodynamic radii and fractal dimensions of the flocs.

Colloid stability is dependent on the balance of the electrostatic repulsive forces and the London-van der Waals attractive forces. This balance is disrupted by the addition of ions and polymers. Addition of ions coagulates the particles by reducing the electrostatic interaction, whereas flocculation is induced by the addition of long chain polymer. Polymer induced flocculation occurs by two main methods: bridging flocculation in which the particles are joined by the same polymer molecule; charge neutralisation occurs when the particle and polymer molecule have opposite charges. Effective flocculants are linear polymers of high molar mass, which may be nonionic but often are polyelectrolytes with numbers of ionizable groups along the chain. The important characteristics of the polymer are its molar mass and charge

0097–6156/93/0532–0161$06.25/0

density. Polyacrylamides are used extensively commercially. Anionic polyacrylamides are prepared by the controlled alkaline hydrolysis of polyacrylamide and find the greatest application as they are available relatively cheaply, however they have the disadvantage that the charge on the polymer chain is pH dependent and they are readily hydrolysed. Cationic polyacrylamides, prepared by the cationic-copolymerisation of acrylamide plus dimethylaminoethyl acrylate followed by quaternisation, are more expensive, but the charge is independent of pH and significant hydrolysis only occurs above pH 10.

In bridging flocculation segments of the same polymer chain adsorb on different particles linking the particles into aggregates. The molecular dimensions of the polymer, dependent on the degree of uncoiling, vary with the pH and ionic strength of the medium, but they are of the order of 0.1 to 1 μm, comparable to the size of the polymer molecule. Excess polymer adsorption can restabilize the particles by steric stabilisation producing a cloudy suspension. In the electrostatic patch model of Kasper and Gregory (1, 2) the particles have a mosaic charge distribution; negative patches are present from the original particle surface, whereas positive patches are produced by the adsorbed cationic polyelectrolyte. Positive and negative patches from different particles attract and bond the particles into an aggregate. The flocculation rate for this type of electrostatic flocculation will decrease with increasing ionic strength of the surrounding medium. It would also be expected to be favoured by shorter chain polymers, which after adsorption do not have sufficient span to bridge to another particle.

Materials and Methods

The kaolinite was purified from matrix stope kaolin supplied by English China Clays. The mineral content was 98% kaolinite and 2% muscovite (XRD) and particle size 1.1 ± 0.1 μm (esd). The surface charge of the kaolinite is negative and increases with pH and ionic strength of added electrolyte. At the constant ionic strength of 10^{-3} mol dm^{-3} used in the flocculation experiments the charge density varies approximately linearly with pH from $- 0.5$ C g^{-1} at pH 3 to $- 2.5$ C g^{-1} at pH 10. BET nitrogen adsorption gave a surface area of 10.5 ± 0.2 m^2 g^{-1}. The titanium dioxide was Fison's SLR grade and was dispersed in water at pH 10 without further treatment. The density of the titania particles was determined to be 3.67 g dm^{-3} by measuring the density of a suspension of known mass fraction. The particles appeared approximately spherical under the electron microscope and the assumption that they approximate to spheres is made throughout. The polystyrene microspheres were monodisperse and obtained from Interfacial Dynamics Corporation; their diameters together with the Smoluchowski rate constants found for aggregation in 1 mol dm^{-3} KCl are given in Table I.

AnalaR sodium and potassium chlorides were used without further purification. All water used was Milli-Q quality and for the light scattering

Table I. Diameters of the polystyrene microspheres and Smoluchowski rate
constants for the monodisperse suspensions in 1 mol dm^{-3} KCl

Diameter, d/nm	Smoluchowski Rate Constant, k_2 x 10^{12} /cm^3 s^{-1}
121	1.37 ± 0.05
264	1.68 ± 0.08
303	2.24 ± 0.10
378	2.06 ± 0.07
623	1.87 ± 0.07

studies was filtered through a 0.4 μm filter. The cationic polyacrylamides
used in these studies were synthesised by Dr M Skinner of Allied Colloids.
They were prepared by copolymerisation of acrylamide and
dimethylaminoethylacrylate quaternised with methyl chloride. Four samples
of copolymer of approximately the same cationic charge density but of
different molar masses were used. The molar masses were determined by
viscometry; the constants of the Mark-Howink equation were k = 3.7 x 10^{-2}
cm^3 g^{-1} and α = 0.66. The intrinsic viscosities of the polymers were
measured in 1 mol dm^{-3} NaCl solution as this suppresses ionisation and
allows the polymer to assume a coiled configuration, so that polymers of
differing ionic characters can be compared directly as the effect of the ionic
groups is virtually nullified. The percentage cationic charge density of each
polymer was determined by displacement of chloride from an anionic
exchange resin. The chloride was determined by potentiometric titration on
an E636 Metrohm titroprocessor. These values agreed with those obtained
by colloid titration with potassium polyvinyl sulphate. The molar mass,
charge density and percentage cationic character (mole of charge per 100 g x
molar mass of quaternised monomer) are given in Table II. Potentiometric
titration showed that the cationic charge density was independent of pH over
the pH range 3-9 in solutions of 10^{-3} to 10^0 mol dm^{-3} NaCl.

Table II. Molar masses, charge densities and % cationic character of the
polyacrylamides used

Polymer	Viscosity average molar mass/10^6 g mol^{-1}	Charge density/ 10^3 C g^{-1}	% Cationic Character
A	1.4	0.16	32
B	3.8	0.16	33
C	7.7	0.15	29
D	10.1	0.15	31

Standard solutions of the polymer were prepared by agitating a mixture of
polymer with about one third of the final required amount of water for 2 to 3
hours. The solutions were left to stand for 20 hours, to facilitate polymer
uncoiling, diluted and stirred gently for 3 hours. The solution was left to

stand for 2 hours prior to use. All solutions were used within 24 hours of preparation. Stock solutions of kaolinite in 10^{-3} mol dm^{-3} NaCl were prepared in 1 dm^3 perspex cylinders, leaving the dispersion for 24 hours to ensure that all particles were completely wetted. The pH of the clay suspensions and polymer solutions were adjusted to the required value (to within ± 0.1 pH unit) by adding small quantities of 1.0 mol dm^{-3} NaOH or HCl. The solutions and the suspension were checked regularly to ensure that the pH was stable throughout the experiment. Flocculation of the clay suspension was carried out in a graduated perspex cuvette placed in a Laser-photosedimentometer assembly (3). A known volume (150 cm^3) of the kaolinite stock suspension was measured into the cuvette and 100 cm^3 of the dilute polymer solution quickly and carefully added. The cuvette was sealed and inverted reproducibly 20 times to ensure complete mixing. For the residual turbidity measurements the cuvette was scanned from the top of the suspension to the base at a set time after the initial mixing. Tests were carried out to check that the polymer uptake was complete under the set experimental conditions; it was also found that the polymer adsorption was irreversible.

The method employed for the determination of trace quantities of cationic polyacrylamide was adapted from that of Horn (4) for polyethylenimine. The detection limit was of the order of 1 part in 10^7. In solution cationic polyacrylamide forms a polymer complex with negatively charged potassium polyvinyl sulphate (KPVS); when the metachromatic dye toluidine blue is present, excess KPVS complexes with the dye producing a colour change from blue to purple as the endpoint is reached. The spectral shift was monitored using a specially made Double Beam Photoelectric Titrator. After flocculation of the clay suspension, the supernatant was transferred to plastic centrifugation tubes and centrifuged at 4000 rpm until all the residual clay particles had settled leaving a crystal clear liquid, which was transferred to the standard perspex titrating cuvette. This method of residual polyacrylamide determination was reproducible to ± 2 %.

The rheological studies were carried out using Bohlin Constant Stress and Constant Strain rheometers. For the particle and floc sizing experiments a Malvern 3600D Fraunhoffer diffraction system was used. For the dynamic light scattering studies the photon correlation system was the Malvern 4700c. The system is homodyne and uses a Coherent Innova 90 laser. A 90° scattering angle and laser wavelength of 488 nm were used. The coagulation experiments were performed by first mixing equal portions of dilute suspensions and 2 mol dm^{-3} potassium chloride solution. The mixture was then transferred to a precision silica optical cell which was then placed in the thermostatted cell-head (25.0 ± 0.1°C) and allowed to equilibrate for 5 minutes. Intensity autocorrelation functions of the coagulating suspensions were then collected at 2 minute intervals allowing 30 s to collect the data for each. The method used to analyze the data is described in ref (5).

Theory

For Newtonian behaviour $\tau = \eta \, \dot{\gamma}$, where τ is the shear stress, η is the viscosity and $\dot{\gamma}$ is the strain rate. Kaolinite dispersions are cohesive sediments and show shear thinning behaviour ie η decreases as $\dot{\gamma}$ increases. The extrapolated value of τ at zero shear rate is called the Bingham yield stress, τ_B. For a Hookean solid $\tau = G \, \gamma$, where G is the modulus of rigidity. Most substances are neither purely elastic (solid-like) nor purely viscous (liquid-like) but are viscoelastic. Kaolinite is in this category. For viscoelastic behaviour

$$\tau = G \, \gamma + \eta \, \dot{\gamma}$$

In the linear viscoelastic region G is constant. By application of a small amplitude oscillatory shear, the storage and loss modulus can be obtained:

$$G^* = G' + iG'',$$

where G is the shear modulus, G' is the storage modulus and G" is the loss modulus. The storage modulus is a measure of elastic or "solid-like" behaviour and the loss modulus is a measure of viscous or "liquid-like" behaviour.

Information on interparticle forces can be obtained from rheological studies. As the shear rate is increased, the particles rearrange, or flocs distort in a flocculated system, and eventually the system starts to flow. The magnitude of the Bingham yield stress, τ_B, reflects: maximum rearrangement of the particles in an unflocculated system; deformation of the flocs in a flocculated system before flow. The shear moduli, G^*, G' and G" effectively probe the unperturbed system at very low shear rates and reflect: the particle-particle interactions in a pure clay suspension; the intraparticle structure of the flocs in a flocculated suspension.

Results and Discussion

The subsidence rate of the kaolinite suspension, flocculated by the cationic polyacrylamides, **of the same charge density,** increased with increasing molar mass of the polymer. This is shown in Figure 1. Thus the polymer of the highest molar mass produces the smallest and most compact flocs. As shown in Figure 2, the supernatant turbidity initially decreases with increasing polymer dosage, but then increases again as the adsorption of excess polymer restabilizes the particles. The two polymers of highest molar mass are most effective at minimising the turbidity and indeed reduce it to zero at a small polymer dosage. However, the adsorption of polymer onto the kaolinite, on a mass to mass basis, increases with decreasing molar mass of the polymer ie the polymer of the lowest molar mass is adsorbed to the greatest extent; the maximum "plateau" values of the polymer adsorption are shown in Figure 3.

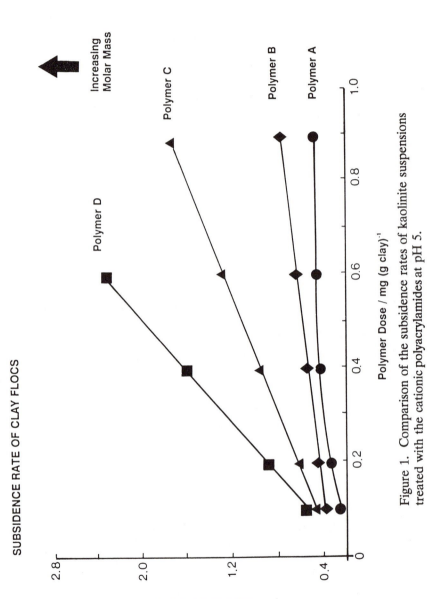

Figure 1. Comparison of the subsidence rates of kaolinite suspensions treated with the cationic polyacrylamides at pH 5.

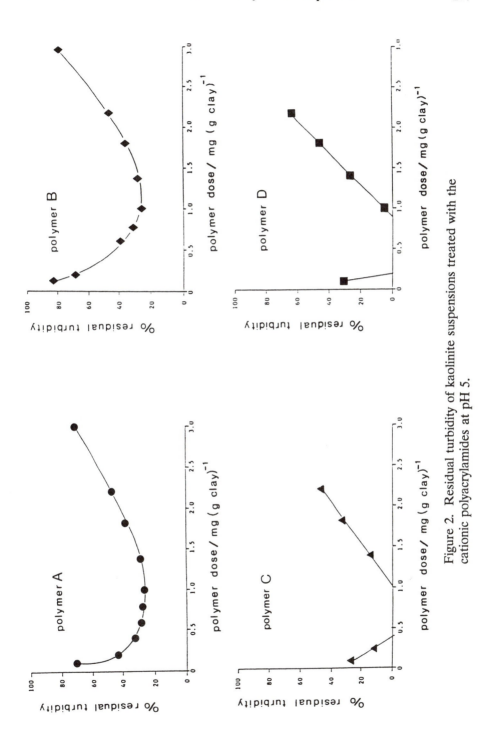

Figure 2. Residual turbidity of kaolinite suspensions treated with the cationic polyacrylamides at pH 5.

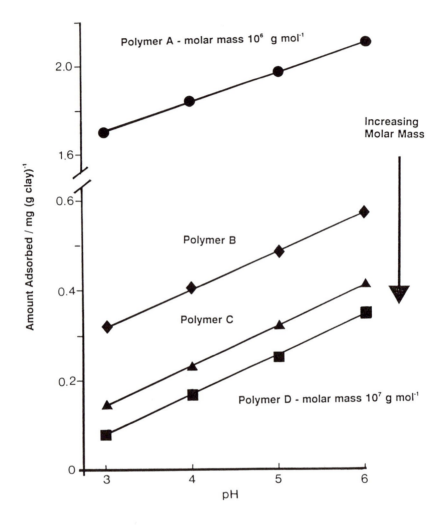

ADSORPTION OF CATIONIC POLYACRYLAMIDES ON KAOLINITE

MAXIMUM "PLATEAU" ADSORPTION VALUES INCREASE WITH PH

Polymer A, of lowest molar mass, adsorbs the most - mass for mass basis.

Figure 3. Maximum "plateau" values of the adsorption isotherms for the cationic polyacrylamides on kaolinite as a function of the pH.

This implies that there is a different mechanism for the bonding of polymers of high and low molar mass. The type of bond will determine the compactness and strength of the flocs. Could rheological studies provide an answer to what determines the type of bond formed and the strength of the flocs under shear?

Kaolinite is a 1:1 layered mineral based on alumina octahedra and silica tetrahedra, with platelets of approximate size 1 μm by 1.5 μm by 50 nm. The face double-layer has an overall negative charge, essentially pH independent; the edge double-layer has a positive charge at pH < 6, but carries a negative charge above this pH (6). It has long been considered that at low values of the pH, when the edge and face have opposite charges, that the kaolinite particles form a "house-of-cards" structure. In the rheological studies of pure kaolinite, the effect of varying the electrolyte concentration and pH were carried out at constant ionic strength. Viscometry studies (Figure 4) on kaolinite suspensions without added polymer, showed that the yield stress decreases with increasing pH. This reflects the decreasing interaction between the particles and their peptisation at high pH. The results of oscillatory measurements for the kaolinite alone are shown in Figure 5. G' is greater than G", confirming that kaolinite behaves as an elastic solid. Both moduli decrease with increasing pH, confirming the decrease in particle-particle interactions. At pH 5, the ionic strength was varied from 10^{-3} to 10^{0} mol dm^{-3} and it was found that the Bingham yield stress decreased with increasing ionic strength; this implies that the "house-of-cards" structure is weakened as repulsion between the kaolinite platelets is reduced.

Several different techniques were used to probe the binding in the flocculated systems. The results of the adsorption studies at pH 5 are given in Figure 6. Polymer A, of lowest molar mass is adsorbed to the greatest extent. The "plateau" values of adsorption were confirmed by viscometry as shown in Figure 7. The viscosity of the system does not increase above the plateau value. This again shows that the rheological properties do indeed reflect the physical chemistry of the system.

Light scattering (Figure 8) confirmed that the flocs formed by polymer D of highest molar mass were the smallest. This is in agreement with the results of the subsidence rate experiments shown in Fig 1. The polymer of highest molar mass forms the densest and most compact flocs. Viscometry studies for the flocculated system at pH 6 are shown in Figure 9. Both the viscosity and yield stress values decrease with increasing molar mass of the polymer, and the values for kaolinite alone are the lowest of all. Thus the flocs formed by polymer A give a more gel-like nature to the system. This is reflected in the subsidence rate studies; the boundary between the flocculated dispersion and supernatant settles as a continuous boundary whereas the boundary from addition of polymer D is granular.

Oscillation studies on the flocculated system were carried out within the linear viscoelastic region. As shown in Figure 10, the storage modulus, G', is greater than the loss modulus, G". Also G' for polymer A is greater than

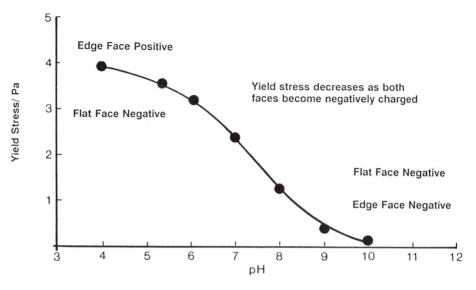

Viscometry Continuous Shear - 120 g dm⁻³ Sodium Kaolinite
                              Ionic Strength - 10⁻³ mol dm⁻³ NaCl

Figure 4. Yield stress as a function of pH for kaolinite alone.

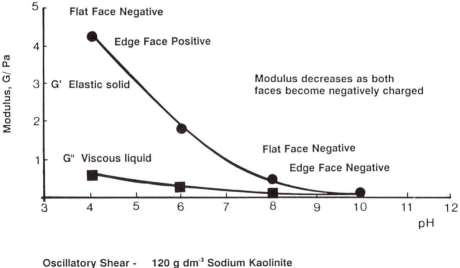

Oscillatory Shear -   120 g dm⁻³ Sodium Kaolinite
                       Ionic Strength - 10⁻³ mol dm⁻³ NaCl
                       ω = 1 Hz

Figure 5. Modulus as a function of pH for kaolinite alone.

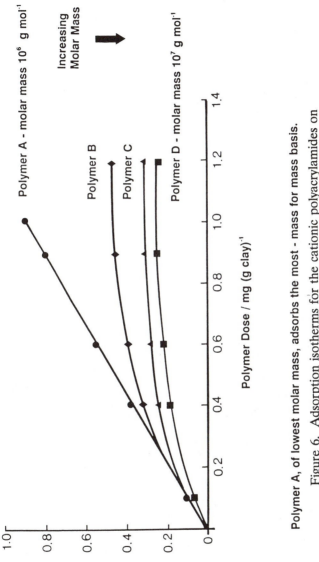

Polymer A, of lowest molar mass, adsorbs the most - mass for mass basis.

Figure 6. Adsorption isotherms for the cationic polyacrylamides on kaolinite at pH 5.

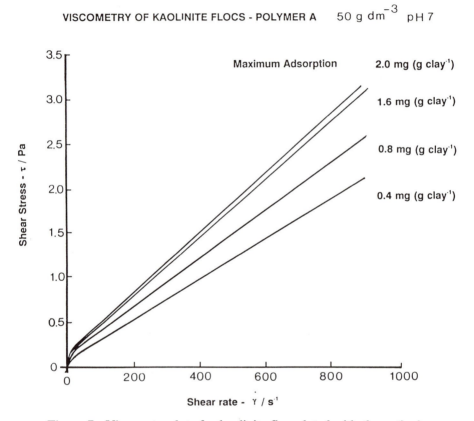

Figure 7. Viscometry data for kaolinite flocculated with the cationic polyacrylamide of lowest molar mass, polymer A.

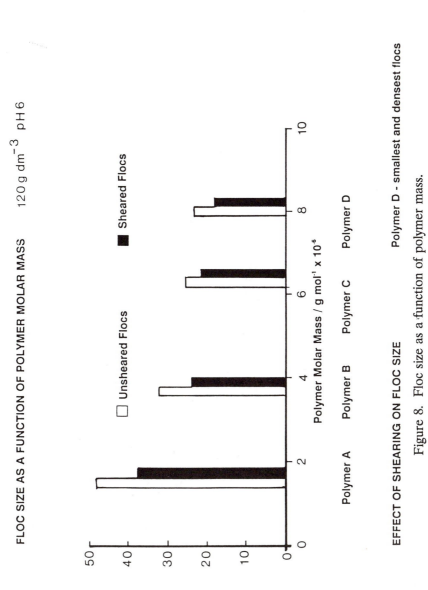

Figure 8. Floc size as a function of polymer mass.

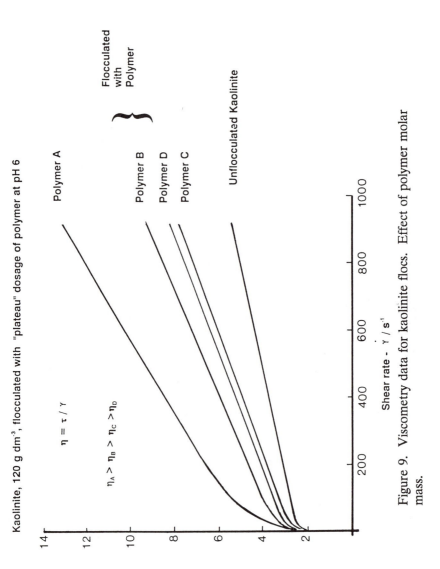

Figure 9. Viscometry data for kaolinite flocs. Effect of polymer molar mass.

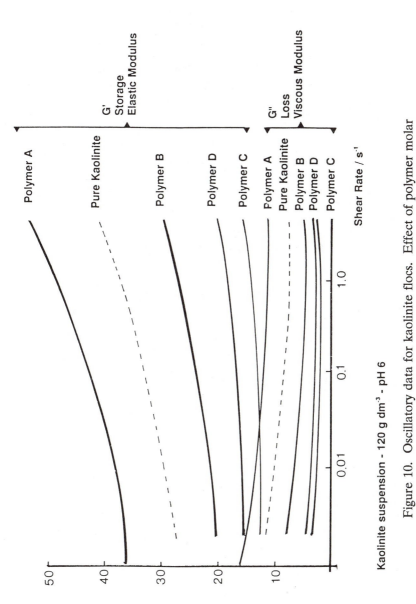

Figure 10. Oscillatory data for kaolinite flocs. Effect of polymer molar mass.

Kaolinite suspension - 120 g dm^{-3} - pH 6

that for kaolinite alone, which in turn is greater than that for polymers B, C and D. The magnitude of the modulus reflects the particle-particle interactions **within** the floc as the system is undergoing a very small perturbation. This shows that the binding within the flocs formed by polymer A is greater than that in kaolinite alone at pH 6, implying that the binding of polymer A is electrostatic in origin.

The effect of pH on the rigidity modulus is given in Figure 11. For polymer A, of lowest molar mass, G **increases** with increasing pH, whereas for polymers B, C and D, the rigidity modulus **decreases** with increasing pH. However, the polymers are all cationic polyelectrolytes of the same charge density and differ only in their molar mass. The surface charge on the clay becomes more negative with increasing pH and the fact that polymer A is bound most strongly supports the electrostatic binding mechanism for this polymer of lowest molar mass.

The electrolyte induced rapid coagulation of polydisperse mixtures of polystyrene microspheres, titania and kaolinite was studied by Photon Correlation Spectroscopy (PCS). A method has been developed that enables the rate constant to be calculated without precise knowledge of the particle size distribution (7). For Brownian Diffusion-controlled Aggregation:

$$E = k_{12} N_0 t$$

where $E = t / t_{1/2}$ is a measure of the extent of aggregation,
    N_0 is the initial number density,
    k_{12} is the second order Smoluchowski rate constant.

For rapid coagulation induced by the addition of 1.0 mol dm^{-3} KCl, the Smoluchowski kinetic scheme gave an excellent description for the rapid coagulation of colloidal dispersions (Figures 12 and 13). This was shown by: the linearity of the second order rate constant plots; the lack of any concentration effects; the closeness of the final rate constants to the diffusion-limited value of 6.1 x 10^{-12} cm^3 s^{-1}.

The structure factor of a coagulating monodisperse spherical suspension has been calculated by computer-simulation methods, assuming cluster-cluster and ballistic particle-cluster aggregation schemes, in both cases subject to the constraint of Smoluchowski kinetics and the Rayleigh-Gans-Debye approximation (8). The simulations were parametrised using theoretical expressions first derived in polymer theory for a 'blob' model together with the concept of fractal packing. Good agreement was found for the radius of gyration and hydrodynamic radius by using a 'blob' size of 1.5 together with the appropriate fractal dimension obtained from the computer simulations (d_f = 1.9 for the cluster-cluster model and 3.0 for the ballistic model).

The variation of the fractal dimension of kaolinite flocs with pH was found using the method of dynamic scaling. Weitz et al (9) showed that for a rapidly aggregating system (a condition fulfilled by kaolinite) that

$$R_z \sim t^{1/d_f},$$

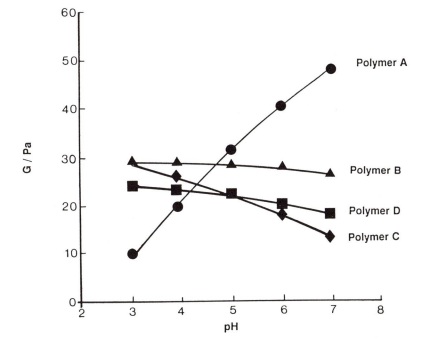

Kaolinite - 120 g dm^{-3} ; plateau polymer dosage; ω = 1 s^{-1}.

Figure 11. Effect of pH on the rigidity modulus of flocculated kaolinite.

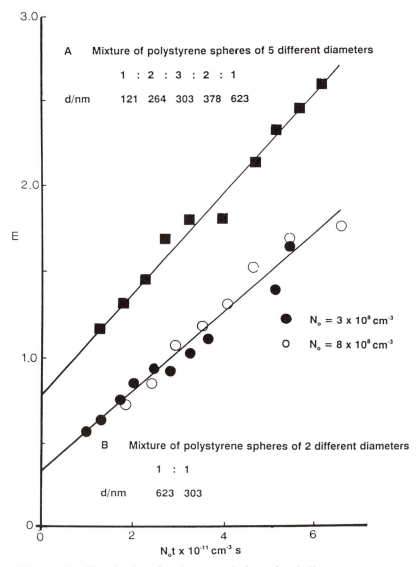

Figure 12. Kinetic data for the coagulation of polydisperse systems.

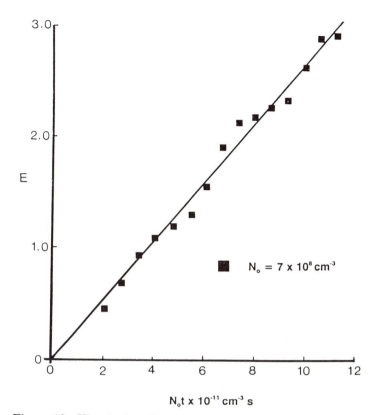

Figure 13. Kinetic data for the coagulation of titanium dioxide suspensions.

where R_z is the harmonic z-average radius of the aggregates in the suspension as determined by PCS and d_f is the fractal dimension of the aggregate. Thus measurement of R_z as time proceeds will yield a log-log plot of R_z against t with slope $1/d_f$. The results are shown in Figure 14. For face-to-face aggregation, $d_f = 3$ and for face-to-edge, $d_f \sim 1.8$. In 1.0 mol dm^{-3} KCl (pH 7.5) $d_f = 3.0$, consistent with a close-packed structure formed by face-to-face aggregation. In water the fractal dimension = 2.7 at pH values below 3 and 1.8 at pH values above 3. In 1 x 10^{-3} mol dm^{-3} KCl the transition from low to high fractal dimension occurs at a higher pH value between 4 and 4.5. Thus increasing the ionic strength at low pH induces face-to-face aggregation. This shows that the effect of the electric double layer is all-important in determining the structure of the system. The effect of the addition of electrolyte is to reduce the face-to-face repulsion so that the compact structure is preferred to the "house-of-cards" open structure to a higher pH value.

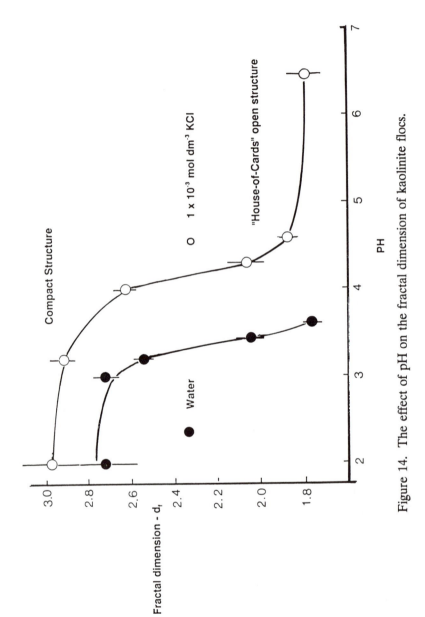

Figure 14. The effect of pH on the fractal dimension of kaolinite flocs.

Conclusions

Rheology has proved to be a powerful technique for determining the types of forces acting between particles in kaolinite dispersions. The interactions are highly dependent on the pH, electrolyte concentration and the nature of the polyelectrolytes present. A series of flocculating suspensions was monitored by dynamic light scattering and the rate constants obtained were in excellent agreement with Smoluchowski theory. The structure factor of the flocs was calculated by computer simulation methods, assuming cluster-cluster and ballistic particle-cluster schemes. The results were interpreted in terms of the fractal dimensions of the flocs. Dilute kaolinite flocs show a transition from loosely-packed voluminous "house-of-cards" type structures in moderate to strong acidic solutions. In concentrated suspensions, however, this gel-like structure is less liable to collapse when face-face repulsions are removed and the gel structure is therefore maintained at low pH.

Acknowledgments

We thank the SERC for their support of this work.

Literature Cited

1 Gregory, J. *J. Colloid Interface Sci.*, **1973**, *42*, 448.
2 Kasper, D. R., California Institute of Technology, PhD Thesis, **1971**.
3 Clarke, A. Q.; Herrington, T. M.; Petzold, J.C. *Colloids and Surfaces*, **1986**, *22*, 51.
4 Horn, D. *Prog. Colloid Polym. Sci.*, **1978**, *65*, 251.
5 Herrington, T. M; Midmore, B. R. *J. Chem. Soc., Far. Trans. I*, **1989**, *85*, 3529.
6 Clarke, A. Q; Herrington, T. M; Watts, J. C. *Colloids and Surfaces*, **1992**, *68*, 161.
7 Herrington, T. M; Midmore, B. R. *J. Chem. Soc., Far. Trans.*, **1986**, *86*, 671.
8 Herrington, T. M; Midmore, B. R; Lips, A. *J. Chem. Soc., Far. Trans.*, **1990**, *86*, 2961.
9 Weitz, D. A; Huang, J. S; Lin, M. Y; Sung, J. *Phys. Rev. Lett.*, **1984**, *52*, 1433.

RECEIVED May 17, 1993

Chapter 15

Interactions of Polyelectrolyte Crystal-Growth Inhibitors with $BaSO_4$ Surfaces

R. G. Thompson

Marathon Oil Company Petroleum Technology Center, 7400 South Broadway, Littleton, CO 80122

Interactions between polyelectrolyte crystal growth inhibitors and $BaSO_4$ crystal surfaces have been studied in an attempt to better understand the factors that control the inhibition of $BaSO_4$ scale growth and deposition in oilfield environments. Modern analytical techniques allow experimental investigation of these systems under more realistic conditions than were previously available. In this investigation, adsorption onto $BaSO_4$, influence on $BaSO_4$ zeta potential, and effectiveness of $BaSO_4$ growth inhibition for poly(acrylic acid) and poly(vinylsulfonic acid) were studied in high ionic strength brines at low pH. These data demonstrate the importance of polyelectrolyte charge density for effective $BaSO_4$ scale growth inhibition under these conditions.

Inhibition of the growth of sparingly soluble mineral crystals such as the carbonate and sulfate salts of Ca^{2+}, Sr^{2+}, and Ba^{2+} is of considerable interest to many industries because of the tendency of these minerals to form scale deposits. In the oil industry, hydrocarbon production is almost always accompanied by production of large volumes of aqueous brine, often leading to the precipitation of mineral salts that form scale deposits. Depending on the nature of the scaling system, deposition can occur within production and processing equipment, causing significant operational problems, or can occur within reservoir rock, causing reduction in permeability and fluid flow. Either occurrence results in significant cost from loss of production and remediation expense. Deposition of sulfate salts of Group IIa cations such as Ba^{2+} can be especially problematic.

Barium sulfate scales form in situations where production of reservoir fluids causes mixing of incompatible aqueous fluids. For example, in North Sea (UK) offshore hydrocarbon production, seawater is injected into reservoirs to displace oil, and maintain reservoir pressure. When seawater, high in sulfate, contacts reservoir fluids that have high concentrations of Ba^{2+}, $BaSO_4$ scales result. Barium sulfate is an especially intractable scale mineral because of its physical hardness and very low solubility (1-6).

Removal of scale deposits can be accomplished either chemically or mechanically, but treatments are expensive and often ineffective. For this reason, prophylactic treatment to inhibit the formation of scale deposits is an attractive approach, especially

0097–6156/93/0532–0182$06.00/0

for BaSO$_4$ scale control. Inhibition of scale formation in subsurface equipment and rock formations is especially important because of the inaccessibility of the downhole environment for remedial action (*1, 2, 5, 6*).

The chemical environment under which BaSO$_4$ scale inhibition must be accomplished is often challenging and not easily controlled. In addition, operational and economic requirements place difficult limits on the chemical systems that can be used. For example, in North Sea oil production, BaSO$_4$ scale inhibition must be accomplished in the subterranean environment. Downhole conditions can be chemically harsh with pressures as high as 7,000 psi, temperatures of 120 °C or higher, brine ionic strength of >1M, pH as low as 4, and very high supersaturation with respect to BaSO$_4$ precipitation. Operational difficulties arise from the need to stop production of a well in order to carry out an inhibition treatment. Scale inhibitors are placed d~~~~' ~ ~ means of a "squeeze" treatment. This treatment method is ⌐rces (*1, 2, 5-10*). In summary, a relatively concentrated ⌐r is placed, or "squeezed", into the rock formation near a ⌐rbed by the rock matrix. Then when the well is returned to molecules desorb from the rock matrix into the produced hibit BaSO$_4$ crystal growth.
itions vary widely, oilfield scale inhibitors for BaSO$_4$ meet the following requirements:
pplication methods, the inhibitor chemistry must have a percent by weight, usually at least 5-10 wt. %.
ignificantly retard crystal growth in produced fluids for an uually several hours.
⌐rmous volume of brines that require treatment in oil ⌐rs must be effective at concentrations well below the ⌐oncentrations of scaling ions in order to be economically uually requires good effectiveness at a concentration at or

⌐main in the aqueous phase of a system containing both gnificant partitioning into the hydrocarbon phase severely ⌐ess.
⌐inate many interesting possibilities for scale inhibition ⌐ful in other crystal growth applications. For example, ⌐obic character can be quite effective for inhibiting crystal ⌐ut are generally not suitable for oilfield applications.
⌐an be met by a class of mineral crystal growth modifiers ⌐ld inhibitors. Threshold inhibitors are surface active olyelectrolytes, that modify mineral crystal growth at ⌐n below stoichiometric concentrations of scaling ions in ⌐4 scale inhibition is usually attempted by treatment of ⌐ctive polyelectrolytes of various chemical functionality,

⌐r BaSO$_4$ scale inhibition include polyphosphonates, ⌐cently, polysulfonates, as well as some copolymers of ⌐nterest for subterranean BaSO$_4$ scale inhibition are ⌐nylsulfonic acid) (*5-17*). Because these molecules are ⌐an produce highly charged species in solution with a ⌐er molecule. Each of these polyelectrolytes has an ⌐tiveness outside of which they become much less ⌐his corresponds approximately to the pH range for ⌐ized.

BaSO$_4$ crystal growth from simple systems is well ⌐⌐⌐⌐u, the influence of growth inhibitors on the growth and deposition process is not. Kinetic studies have shown that BaSO$_4$ crystal growth is surface controlled, and is thought to proceed by a spiral growth mechanism where new growth edges are continuously generated by the emergence of screw dislocations in the crystal (*14-17*). Threshold growth inhibitors are thought to interfere with this growth process by adsorbing at active surface growth sites such as kinks and ledges. Adsorption of inhibitors at active surface growth sites creates both a steric and electrostatic barrier to further crystal growth. Thus, development of the growing crystal lattice can be retarded until such time as the adsorbed inhibitor is overgrown, or until the crystal surface develops new, exposed growth sites.

In an attempt to better understand the factors that control the inhibition of $BaSO_4$ crystal growth and deposition in an oilfield environment, we have conducted studies of the interactions between polyelectrolyte growth inhibitors and $BaSO_4$ crystal surfaces. Recent advances in analytical techniques have allowed experimental investigation of polyelectrolyte adsorption at low concentrations, as well as accurate, reliable determination of the effect of polyelectrolyte adsorption on $BaSO_4$ zeta potential, a sensitive probe of polyelectrolyte interaction with $BaSO_4$ surfaces. Correlation of adsorption and zeta potential results with the performance of polyelectrolytes as $BaSO_4$ growth inhibitors has yielded useful insight into the mechanism of scale inhibition in systems of practical interest.

The polyelectrolytes investigated in the present study are a poly(acrylic acid) of molecular weight(wt. avg.) 15,000, and a poly(vinylsulfonic acid) of molecular weight(wt. avg.) 18,000. The conditions used in these experiments are relevant to those encountered in the subterranean environment of the North Sea (UK) where $BaSO_4$ scale is a problem. Solution conditions include high ionic strength brines of 1M NaCl, low solution pH of 4, and low polyelectrolyte scale inhibitor concentrations.

For simplicity in the discussion that follows, poly(acrylic acid) will be referred to hereafter as PAA, and poly(vinylsulfonic acid) will be referred to as PVS.

Experimental

Adsorption Experiments. $BaSO_4$ was purchased from Aldrich (99%). Before use, the $LaSO_4$ was washed three times with deionized water to remove any soluble salt impurities, oven-dried at 120 °C for at least 24 hours, and stored at 120 °C. Scanning electron micrographs showed the $BaSO_4$ crystals to be rounded rhombs of approximately 0.3-0.5 μm in diameter. The $BaSO_4$ had a surface area of 2.3 m^2/g determined by B.E.T. method nitrogen adsorption analysis on a Micromeritics AccuSorb 2100E surface area analyzer.

ACS reagent grade NaCl was purchased from Mallinckrodt. Standard HCl and NaOH solutions were purchased from Mallinckrodt in concentrations of 0.01M and 0.1M.

The polyelectrolyte scale inhibitors used were a poly(acrylic acid), purchased from B. F. Goodrich with a molecular weight(wt. avg.) of 15,000, and a poly(vinylsulfonic acid), with a molecular weight(wt. avg.) of 18,000. The PVS was prepared by published methods (*18*).

Polyelectrolyte solutions were prepared in 1.0 M NaCl from solid samples dried to constant weight. Solutions for adsorption experiments ranged in concentration from 10-25 ppm of polyelectrolyte. The pH of the polyelectrolyte solutions was adjusted to the desired value with standard HCl or NaOH as needed and allowed to equilibrate overnight prior to adsorption experiments.

Adsorption experiments were carried out by placing 10 ml of aqueous polyelectrolyte solution of known concentration in a capped polypropylene test tube along with a carefully weighed amount of $BaSO_4$. Ratios of polyelectrolyte to $BaSO_4$ were controlled to give initial values from approximately 0.1 mg polyelectrolyte/m^2 $BaSO_4$ to 10 mg polyelectrolyte/m^2 $BaSO_4$. The resulting mixtures were shaken to assure complete suspension of BaSO$_4$, placed in a thermostated bath, and shaken periodically to insure intimate contact between polyelectrolyte solution and BaSO$_4$ surfaces. The suspensions were allowed to equilibrate for 24 hours, although adsorption kinetics experiments showed that essentially all adsorption was complete after at most two hours.

Samples were then withdrawn and filtered through 0.22 μm syringe filters (Alltech Anotop 10) to remove all BaSO$_4$ solids. Equilibrium polyelectrolyte solution concentrations were determined by HPLC size exclusion chromatography using a Shodex Q801 (500 mm x 8 mm) 420 nm aqueous size exclusion column. This method has a detection limit of about 0.7 ppm and a precision of better than ± 3% over the range of concentrations of these samples. Control experiments were done to verify that adsorption of polyelectrolyte by polypropylene containers, or by contact with syringes, filters and vials used during sample handling was not detectable under the conditions of these experiments.

Zeta Potential Determinations. The effect of polyelectrolyte scale inhibitor concentration on BaSO$_4$ zeta potential was investigated by titrating BaSO$_4$ suspensions with solutions of polyelectrolyte and determining BaSO$_4$ zeta potential at increasing concentration of polyelectrolyte. BaSO$_4$ suspensions were prepared from 75 mg of dry BaSO$_4$ (surface area = 2.3 m^2/g) in 250 ml of 1.0 M NaCl. The suspensions were sonicated for 10 minutes to facilitate suspension and the initial pH of the suspensions was adjusted to the desired value with standard HCl or NaOH as needed and maintained at a constant value throughout the titration. The suspensions were stirred vigorously overnight prior to titration with polyelectrolyte.

Titrant solutions were prepared with 500 ppm concentrations of polyelectrolyte in 1.0 M NaCl. The pH of the titrants was adjusted with either HCl or NaOH as needed.

The BaSO$_4$ suspension was then titrated stepwise with polyelectrolyte and allowed to equilibrate for about 30 minutes after each addition. The zeta potential of the resulting suspension was then determined by Doppler electrophoretic light scattering analysis using a Coulter Electronics DELSA 440. BaSO$_4$ suspensions were titrated to a total polyelectrolyte concentration of 100 mg polyelectrolyte/m^2 BaSO$_4$.

Polyelectrolyte Inhibitor Performance. The effect of polyelectrolyte concentration on BaSO$_4$ inhibition performance was determined using a testing method proprietary to Marathon Oil Company (*Thompson, R. G., Marathon Oil Company, internal report*). In this test, the ability of the polyelectrolyte to inhibit the seeded growth of BaSO$_4$ from synthetic brines simulating actual field conditions was determined. Results are phrased in terms of an inhibition efficiency expressed as the percentage of initial Ba^{2+} remaining in solution at the end of the test time.

ACS reagent grade MgCl$_2$·6H$_2$O, CaCl$_2$·2H$_2$O, SrCl$_2$·6H$_2$O, BaCl$_2$·2H$_2$O, and Na$_2$SO$_4$ for these experiments were purchased from Mallinckrodt.

Results and Discussion

Adsorption of Polyelectrolyte onto BaSO$_4$. Determination of experimental adsorption isotherm data for BaSO$_4$ scale inhibitors is important because adsorption of polyelectrolyte inhibitor molecules at active surface growth sites is a key step in crystal growth inhibition (*16, 17, 19-25*). The conditions under which adsorption experiments are carried out are very important because factors such as ionic strength and pH strongly influence both the characteristics of polyelectrolytes and the characteristics of the BaSO$_4$ surface. Adsorption isotherms for PAA adsorption onto BaSO$_4$ have been previously reported, but were determined under conditions of lower ionic strength and higher pH than is of interest in the present study. In addition, higher PAA molecular weight and the presence of spin labels on the PAA molecule distinguish previous studies from the present one (*26-31*).

Interestingly, while the adsorption of polyelectrolytes onto $BaSO_4$ does not conform to the usual assumptions of the Langmuir adsorption isotherm model, the Langmuir model nevertheless provides a good fit to these experimental results as it does in many other cases of polymer adsorption. The Langmuir adsorption isotherm can be written in linear form as shown in Equation 1.

$$C/Q = 1/KN + C/N \qquad (1)$$

Where

Q = equilibrium surface concentration of adsorbed species,
C = equilibrium solution concentration of adsorbing species,
K = relative adsorption affinity constant,
N = surface saturation concentration.

The values of K and N can be found from the slope and intercept of a plot of C/Q vs. C. The exact physical significance of these constants for this system is debatable. Nonetheless, they are useful for purposes of comparison between experiments done under equivalent conditions (*32, 33*).

Figure 1 shows adsorption isotherms for adsorption of PAA and PVS onto $BaSO_4$ in 1.0 M NaCl at pH = 4, conditions that model North Sea subterranean fluid conditions. The figure clearly shows that at pH = 4, PAA is adsorbed at the $BaSO_4$ surface to a much greater extent than is PVS. Both the relative adsorption affinity K (mg Polymer/m^2 $BaSO_4$)$^{-1}$, and the surface saturation concentration N (mg Polymer/m^2 $BaSO_4$), are greater for PAA than for PVS. The adsorption isotherm for PAA under these conditions is similar to that reported by Cafe and Robb for adsorption of a radio-labelled PAA onto $BaSO_4$ at a somewhat higher pH of 4.7, and a lower ionic strength of 0.5 M NaCl (*26*). It must be noted that the values of N for both of these polyelectrolytes exceed by a factor of 2-3 the value that would be calculated for 100% surface coverage. This implies that the configuration of these polymers on the surface is not flat, but rather involves many loops and tails containing unadsorbed functional groups extending away from the surface. Such loops and tails are stabilized by the high ionic strength 1.0 M NaCl bulk liquid phase.

For PAA and PVS, the principle factor influencing adsorption under these conditions is the degree of ionization of the polymers. Figure 2 shows the percent ionization of PAA (pK_a = 4.5) and PVS (pK_a <2) across a wide range of pH. As shown in the figure, PAA is only 24% ionized at pH = 4, while PVS is essentially 100% ionized. Therefore, the charge density on PVS is much greater than that on PAA at pH = 4. This difference has several ramifications in terms of electrostatic interactions and polyelectrolyte solubility.

The principle electrostatic effect is the repulsion that arises between adjacent charged polymer segments adsorbed onto the $BaSO_4$ surface. The repulsion arises between charged groups that are specifically adsorbed at the surface and between those that extend away from the surface in loops and tails, but are in proximity to each other. It has been shown that this repulsion is decreased in the presence of electrolyte (*34, 35*). Still it is much greater for PVS than for PAA because of the greater charge density of PVS at pH = 4. This favors the adsorption of PAA compared to that of PVS.

Another result of the difference in percent ionization between PAA and PVS at pH = 4 is that 1.0 M NaCl is a poorer solvent for PAA than it is for PVS. This provides an appreciably greater solubility drive for PAA adsorption onto the $BaSO_4$ surface than exists for PVS.

These effects favor the adsorption of PAA onto $BaSO_4$ compared to the adsorption of PVS which is a necessary step in the inhibition of $BaSO_4$ crystal

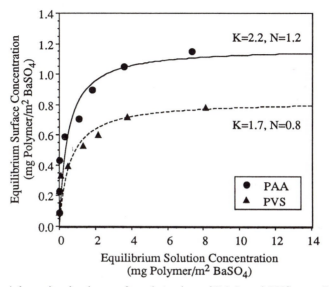

Figure 1. Adsorption isotherms for adsorption of PAA and PVS onto BaSO$_4$ in 1.0 M NaCl at pH=4.

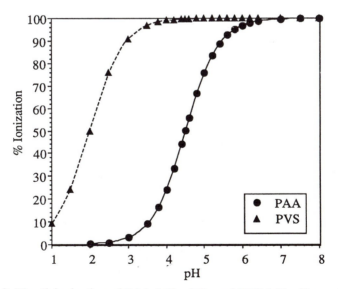

Figure 2. The % ionization of PAA (pK$_a$=4.5), and PVS (pK$_a$=2) as a function of system pH.

growth. However, the data presented below show that this is not the only property needed for effective $BaSO_4$ growth inhibition under these conditions.

Influence of Adsorbed Polyelectrolyte on $BaSO_4$ Zeta Potential.
While some literature reports of zeta potential determinations for $BaSO_4$ systems exist, many of the reported results are problematic for one reason or another (36-47). Several reports involve experiments where ionic strength and pH were not adequately controlled, and few measurements were obtained under conditions of practical interest.

Recent development of laser-based instrumentation for electrophoretic mobility determination has made it possible to determine zeta potential of particles suspended in liquid media for systems that were difficult or impossible to study by classical techniques. These instruments measure electrophoretic mobilities by making direct velocity measurements of particles moving in an applied electric field by analyzing the Doppler shift of laser light scattered from the moving particles. Electrophoretic mobilities can then be converted into zeta potentials by use of standard equations (48-50).

Figure 3 shows the effect of polyelectrolyte concentration on $BaSO_4$ zeta potential for both PAA and PVS. Under these conditions, the zeta potential of $BaSO_4$ in the absence of any polyelectrolyte is about +3 mV. However, both polyelectrolytes cause the $BaSO_4$ zeta potential to become much more negative even at very low concentrations. With increasing concentration of PAA, $BaSO_4$ zeta potential becomes more negative until it reaches a plateau value of -5 mV above a PAA concentration of about 5 mg/m^2. With increasing concentration of PVS, $BaSO_4$ zeta potential becomes more negative until it reaches a plateau value of -17 mV above a PVS concentration of about 2 mg/m^2.

The difference in the magnitude of the effects of PAA and PVS on $BaSO_4$ zeta potential arises from the difference in % ionization of the two polyelectrolytes shown in Figure 2. Even though PAA is adsorbed onto $BaSO_4$ to a greater extent than is PVS, the greater degree of ionization of PVS means that PVS delivers more negative charge density to the $BaSO_4$ surface, causing a much more negative $BaSO_4$ zeta potential than does PAA. Also, because the surface saturation concentration is lower for PVS than for PAA (Figure 1), PVS achieves its maximum influence on $BaSO_4$ zeta potential at a lower concentration than does PAA. This has important implications for $BaSO_4$ growth inhibition.

The negative zeta potential imparted to the $BaSO_4$ surface by adsorption of PAA or PVS is important because it establishes an electrostatic repulsion to further adsorption of charged species, including negatively charged lattice ions (35). This repulsion of lattice ions can produce a large kinetic barrier to crystal growth. The negative $BaSO_4$ zeta potential also stabilizes suspended $BaSO_4$ particles against aggregation.

$BaSO_4$ Inhibition Efficiency. A comparison of the effectiveness of PAA and PVS in $BaSO_4$ scale inhibition is shown in Figure 4. The results shown in Figure 4 were obtained with a testing procedure that determines the ability of a polyelectrolyte scale inhibitor to inhibit the seeded growth of $BaSO_4$. Solution conditions were set to simulate real field conditions of interest as closely as possible. In this case, the test conditions model those encountered in the downhole environment of the North Sea (UK) Brae Field. High ionic strength brines supersaturated with respect to scaling minerals, especially $BaSO_4$, were used at pH=4. The % Efficiency in Figure 4 is the percentage of initial Ba^{2+} remaining in solution at the end of one hour.

It is clear from Figure 4 that PVS is significantly more effective than PAA at inhibiting $BaSO_4$ scale growth under these conditions. At all polyelectrolyte concentrations shown, PVS is more effective than PAA.

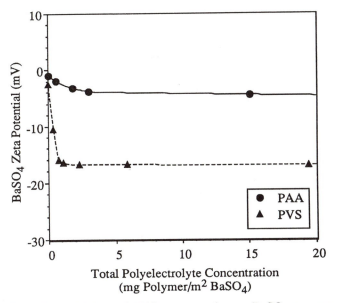

Figure 3. The effect of PAA and PVS concentration on BaSO$_4$ zeta potential in 1.0 M NaCl at pH=4.

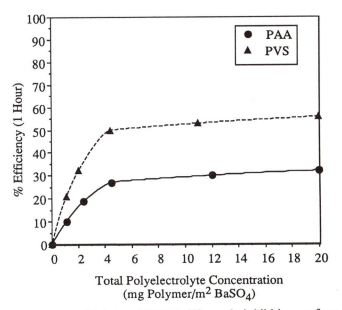

Figure 4. Comparison of PAA and PVS BaSO$_4$ scale inhibition performance at pH=4 after one hour.

These results are somewhat surprising in view of the adsorption results shown in Figure 1. Even though PAA adsorbs onto $BaSO_4$ to a greater extent than does PVS under these conditions, PVS adsorption results in a much more negative $BaSO_4$ zeta potential. The greater ability of PVS to deliver negative charge density to a growing $BaSO_4$ surface under these conditions is the key to its inhibition effectiveness compared to PAA. The importance of crystal growth inhibitor charge density has been noted in other systems as well (51-53).

The negative charge density generated on a growing $BaSO_4$ crystal surface by PVS adsorption inhibits $BaSO_4$ growth by generating an electrostatic kinetic barrier to adsorption of SO_4^{2-} lattice ions. In addition, the negative $BaSO_4$ zeta potential impedes scale deposition processes by inhibiting aggregation of $BaSO_4$ particles or deposition onto other surfaces present in the system. This is very helpful in field inhibition applications because $BaSO_4$ particles that form downhole can be stabilized in suspension long enough to be swept through a production system without depositing as scale.

It should be emphasized that the differences reported here in $BaSO_4$ scale inhibition performance between PAA and PVS are specific to the low pH conditions of this study, which are representative of certain conditions encountered in oil production. Under different conditions, specifically higher pH, where the degree of ionization of PAA and PVS are similar, PAA can be as effective if not more so than PVS for $BaSO_4$ scale inhibition.

Conclusions

In high ionic strength, low pH fluids such as can be encountered in downhole oilfield environments, the effectiveness of polyelectrolyte $BaSO_4$ growth inhibitors depends on their ability to adsorb at the growing crystal surface and thereby to effect several changes. The first important effect of polyelectrolyte adsorption is to sterically block surface growth sites, but this is not in itself sufficient for good inhibition. In addition, the polyelectrolyte growth inhibitor must deliver significant charge density to the growth surface. The benefit of the charge density is two-fold. First, an electrostatic kinetic barrier is generated that impedes adsorption of lattice ions of like charge, thus retarding crystal growth. Second, the charge on suspended crystals stabilizes them against aggregation with each other or with other surfaces in the system, thus preventing formation of scale deposits.

In order to provide charge density at a $BaSO_4$ surface, the adsorbed polyelectrolyte must be significantly ionized. Under the pH = 4 conditions of interest in this study, PVS is essentially 100% ionized while PAA is only about 24% ionized. For this reason, PVS is a much more effective $BaSO_4$ growth inhibitor than is PAA at these conditions.

Acknowledgment

The author wishes to acknowledge the support of Marathon Oil Company for the research reported in this article.

Literature Cited

1   Boyle, M. J.; Mitchell, R. W. *Offshore Europe-79, Soc. Petroleum Eng. Paper No. 8164*, Aberdeen, Scotland, **1979**.
2   Mitchell, R. W.; Grist, D. M.; Boyle, M. J. *J. Petroleum Tech.*, **1980**, pp. 904-912..
3   Mitchell, R. W. *J Pet. Tech.*, **1978**, p. 877-884.
4   Weintritt, D. J.; Cowan, J. C. *J. Pet. Tech.*, **1967**, pp. 1381-1394.

5   Cushner, M. C.; Przybylinski, J. L; Ruggeri, J. W. *NACE Corrosion-88, Paper No. 428.*, **1988**.
6   Ray, J. M.; Fielder, M. in *Chemicals in the Oil Industry*, Ogden, P. H. Ed.; Royal Society of Chemistry: London, **1988**, pp. 87-107.
7   Cowan, J. C.; Weintritt, D. J. *Water-Formed Scale Deposits*; Gulf Publishing Company: Houston, TX, **1976**, pp. 250-424.
8   Burr, B. J.; Howe, T. M.; Goulding, J. *Soc. Petroleum Eng. Paper No. 16261*: San Antonio, TX, **1987**.
9   Pennington, J. in *Chemicals in the Oil Industry*, Ogden, P. H. Ed.; Royal Society of Chemistry: London, **1988**, pp. 108-120.
10 van Rosmalen, G. M. *75th Annual Meeting, Precipitation Processes Session, AIChE*, New York, NY, **1981**.
11 van der Leeden, M. C.; van Rosmalen, G. M. in *Chemicals in the Oil Industry*, Ogden, P. H. Ed.; Royal Society of Chemistry: London, **1988**, pp. 67-86.
12 Liu, S. T.; Nancollas, G. H. *SPE Journal*, **1975**, *259*, pp. 509-516.
13 Vetter, O. J. *J. Petroleum Tech.*, **1972**. pp. 997-1006.
14 Leung, W. H.; Nancollas, G. H. *J. Crystal Growth*, **1978**, *44*, pp. 163-167.
15 Vere, A. W. *Crystal Growth. Principles and Progress*, Plenum Press, New York, **1987**, pp. 5-66.
16 Gardner, G. L; Nancollas, G. H. *J Phys. Chem.*, **1983**, *87*, pp. 4699-4703.
17 Leung, W. H.; Nancollas, G. H. *J. Inorg. Nucl. Chem.*, **1978**, *40*, pp. 1871-1875.
18 Falk, D. O.; Dormish, F. L.; Beazley, P. M.; Thompson, R. G. *U.S. Patent 5,092,404*, **1992**.
19 Nancollas, G. H.; Reddy, M. M. *SPE Journal*, **1974**, pp. 117-.
20 Nancollas, G. H. *Adv. in Colloid and Interface Science*, **1979**,*10*, pp. 215-252.
21 Amjad, Z. *J. Colloid and Interface Science*, **1987**, *117(1)*, pp. 98-103.
22 Wat, R. M. S.; Sorbie, K. S.; Todd, A. C.; Chen, P.; Jiang, P. *Soc. Petroleum Eng. Paper No. 23814*, Lafayette, LA, **1992**.
23 van der Leeden, M.C.; van Rosmalen, G. M. *SPE Production Engineering*, **1990**, pp. 467-470.
24 Amjad, Z *Langmuir*, **1987**,*3*, pp. 1063-1069.
25 Liu, S. T.; Griffiths, D. W. *International Symposium on Oilfield and Geothermal Chemistry, Soc. Petroleum Eng. Paper No. 7863*," Richardson, TX, **1979**.
26 Cafe, M.; Robb, I. D. *J. Colloid and Interface Science*, **1982**, *86(2)*, pp. 411-421.
27 Bain, D. R.; Cafe, M. C.; Robb, I. D.; Williams, P. A. *J. Colloid and Interface Sci.*, **1982**, *88(2)*, pp. 467-470.
28 Williams, P. A.; Harrop, R.; Phillips, G. O.; Robb, I. D.; Pass, G. in *Effect of Polymers on Dispersion Properties: Proceedings of an International Symposium*, Tadras, T. F., Ed.; Academic Press, London **1982**, pp. 361-377.
29 Williams, P. A.; Harrop, R.; Phillips, G. O.; Robb, I. D. *Ind. Eng. Chem. Prod. Res. Dev.*, **1982**,*21*, pp. 349-352.
30 Williams, P. A.; Harrop, R. *J. Chem. Soc.*, Faraday *Trans. 1*, **1985**, *81*, pp. 2635-2646.
31 Wright, J. A.; Harrop, R.; Williams, P. A.; Pass, G.; Robb, I. D. *Colloids Surf.*, **1987**, *24 (2-3)*, pp. 249-258.
32 Heimenz, P. C. *Principles of Colloid and Surface Chemistry*, 2nd Ed.; Marcel Dekker, Inc.: New York, NY, **1986**, pp. 398-412.

33   Adamson, A. W. *Physical Chemistry of Surfaces, 4th Ed.*; John Wiley & Sons: New York, NY, **1982**, pp. 370-382.
34   Robb, I. D.; Sharples; M. *J. Colloid and Interface Science*, **1982**,*89(2)*, pp. 301-308.
35   Adam, U. S.; Robb, I. D. *J. Chem. Soc., Faraday Trans. 1*, **1983**, *79*, pp. 2745-2753.
36   Buchanan, A. S.; Heymann, E. *Proc. Roy. Soc.: London*, **1948**, *A195*, p. 150
37   Buchanan, A. S.; Heymann, E. *Nature*, **1948**, *162*, p. 649.
38   Buchanan, A. S.; Heymann, E. *Nature*, **1948**, *162*, p. 691.
39   Buchanan, A. S.; Heymann, E. *Nature*, **1949**, *164*, p. 29.
40   Buchanan, A. S.; Heymann, E. *J. Colloid Sci.*, **1949**, *4*, pp. 137-150.
41   Ruyssen, R.; Loos, R. *Nature*, **1948**, *162*, p. 741.
42   Morimoto, T. *J. Colloid Sci.*, **1964**, *37(3)*, pp. 387-392.
43   Spitsyn, I.; Torchenkova, E. A.; Glazkov, I. N. *Doklady Akademii Nauk SSSR*, **1969**, *187(6)*, pp. 1335-1338.
44   James, A. M.; Goddard, G. H. *Pharmaceutica Acta Helv.*, **1971**, *46*, pp. 709-720.
45   Kim, J. Y. *J. Korean Inst. Mining*, **1970**, *7 (2)*, pp. 60-67.
46   Coffey, M. D.; Lauzon, R. V. *Intnl. Symp. Oilfield Chemistry, Soc. Petroleum Eng. Paper No. 5302*, Dallas, **1975**, pp. 93-96.
47   Yousef, A. A.; Bibawy, T. A. *Tenside Detergents*, **1976**, *13*, pp. 316-321.
48   Hunter, R. J. *Zeta Potential in Colloid Science*, Academic Press, New York, NY, **1981**, pp. 59-124.
49   Shaw, D. J. *Introduction to Colloid and Surface Chemistry*, Butterworths, London, **1980**.
50   Oja, T.; Bott, S.; Sugrue, S. *Doppler Electrophoretic Light Scattering Analysis Using the Coulter DELSA 440*, Coulter Electronics, **1988**, Part no. 4206268-5A 4-88.
51   McCarthey, E. R.; Alexander, A. E. *J Colloid Sci.*, **1958**, *13*, pp. 383-396.
52   Smith, B. R.; Alexander, A. E. *J. Colloid Interface Sci.*, **1970**, *34(1)*, pp. 81-90.
53   Griffiths, D. W.; Roberts, S. D.; Liu, S. T. *International Symposium on Oilfield and Geothermal Chemistry, Soc. Petroleum Eng. Paper No. 7862*, Richardson, TX, **1979**.

RECEIVED February 23, 1993

INTERACTIONS WITH MICELLES, BILAYERS, LIPOSOMES, AND PROTEINS

Chapter 16

Interaction between Sodium Dodecyl Sulfate and Poly(ethylene oxide) in Aqueous Systems

Light-Scattering and Dynamic Fluorescence-Quenching Studies of the Effect of Added Salt

Jan van Stam, Wyn Brown[1], Johan Fundin, Mats Almgren, and Cecilia Lindblad

Department of Physical Chemistry, University of Uppsala, Box 532, S−751 21 Uppsala, Sweden

Light scattering and dynamic fluorescence methods were combined to study the effect of added simple salt on the interaction between the anionic surfactant sodium dodecylsulfate (SDS) and the nonionic polymer poly(ethylene oxide) (PEO).

From dynamic fluorescence quenching (DFQ) measurements surfactant aggregation numbers and fluorescence quenching rates were obtained. The aggregates formed at the onset of interaction are small and increase in size with increasing surfactant concentration, except for the highest salt concentration. At the highest salt concentration, elongated prolate particles or rods are formed with higher aggregation numbers, rather than the small, spherical aggregates typifying low salt conditions.

Diffusion coefficients from dynamic light scattering (DLS) reflect changes in the pair interaction potential as the SDS concentration and ionic strength are altered. Laplace inversion of the DLS time correlation function gives relaxation time distributions consisting of the SDS/PEO-complex as the main component and free SDS micelles as the minor component. The shift in peak areas shows the increase in the degree of binding between SDS and PEO as the relative concentrations change and also the increase in SDS micellar size with increase in the ionic strength of the medium.

Polymer-surfactant systems have found wide-spread practical applications, e.g., in paints, in pharmaceutical formulations, and in systems for enhanced oil recovery. Their practical importance and the fundamental intricacies in these systems have triggered extensive studies of the interactions between polymers and surfactants, and several reviews have appeared [1-6]. The sodium dodecylsulfate (SDS)-poly(ethylene oxide) (PEO) system has been particularily well studied, both by classical [7,8] and modern methods [9-12].

The "phase diagram" presented by Cabane and Duplessix [9b] has had great impact on our thinking regarding surfactant-polymer interactions in general, and on SDS-PEO interactions in particular. This phase diagram is based on observations of two critical concentrations, x_1 for the onset of cooperative interactions between SDS and the polymer, and x_2 for the saturation of the polymer; a third critical concentration x_2' is

---

[1]Corresponding author

suggested to indicate the formation of free micelles. The first critical point, also called the critical aggregation concentration, cac, is almost independent of the polymer concentration and is substantially lower than the cmc of the surfactant alone. The ‾ ‾' ‾‾‾ ‾‾ is constant and equal to the difference between the cmc

is saturated at x_2 the free SDS concentration is

lue, and free SDS micelles form first after additional

Although the experimental support of the detailed

t factor behind its construction is the assumption,

is experimental observations, that at saturation PEO

d composition, so that the difference x_2-x_1 should

tration of PEO. At very low PEO concentrations the

oth x_1 and x_2 must approach the cmc at sufficiently

Duplessix [9b] on the salt dependence of the critical

o outline changes in the diagram with ionic strength,

data are sparse, it is obvious that the ionic strength

equence is that the diagram with no added salt is in

reases the concentration of free counterions in the

counterions remain free when aggregates or micelles

f the electrostatic effect at a level similar to added salt

oncentration. This would mostly affect the salt-free

tant and polymer concentration. 30 mM SDS (8.7

hetical diagram at zero ionic strength, saturates 0.4%

iic strength of 10 mM, more SDS is taken up by the

) would be saturated in reality. The value reported by

3% PEO is actually close to 30 mM SDS.

and Duplessix [9b] focused on the properties of the

The structure is composed of micelle-like aggregates

er at the interface. The aggregates are slightly smaller

t concentrations, and remain small even at high salt

celles grow into long rods. The SDS/PEO-complex – at

I – behaves as a flexible polymer in a good solvent. The

to scale with the molecular weight of the PEO chain as

itself in water or salt, and for the saturated SDS/PEO-

iing [12] and excimer formation [10] have been used to

nbers of the surfactant aggregates on the polymer chain,

and close to the onset of aggregation. The aggregates

t grow with increasing surfactant concentration to sizes

: micelles in water for the saturated chain.

if the polymer-surfactant complex as they have emerged

Cabane et al. [9] and subsequently those of others seem

are certain aspects worthy of a more detailed study. We

have investigated the properties of the polymer-surfactant complex away from saturation, and at lower ionic strengths than 500 mM, using light scattering methods and dynamic fluorescence quenching.

MATERIALS AND METHODS

Materials. PEO designated SE-150 ($M_w=9.96\cdot10^5$; $M_w/M_n =1.12$) from Toya

Soda Ltd. was used for all measurements but the fluorescence quenching measurements on high salt, 500 mM NaCl, solutions and for the static fluorescence measurements. For the 500 mM NaCl solutions PEO designated Polyox WSR 301 (M_w=1.5·10^6) was used and for the static fluorescence measurements PEO with M_w=6·10^5 from BDH was used. Pyrene (Aldrich) and 2,5-di-tert-butylaniline (DTBA) (Aldrich 99%) were recrystallized from ethanol. SDS was from BDH, U.K., and had a cmc in agreement with literature values (≈8 mM) and was used without further purification. All solutions were prepared with distilled water. For light scattering measurements, solutions were filtered through 0.22 μm Millipore (Fluoropore) "Millex" filters directly into the cylindrical light scattering cells (15 mm in diameter), which were then sealed. All measurements were performed at 25 °C.

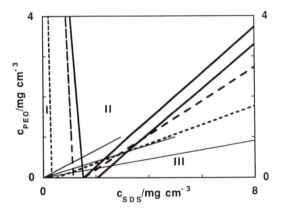

Figure 1. Phase diagram for theSDS/PEO-system adopted from reference 1. Fulldrawn thick lines refer to systems without added salt, slightly broken lines refer to systems with 10 mM salt added, and broken lines to systems with 100 mM added salt. Thin lines refer to SDS/PEO-ratios from top to bottom; 1:3, 1:5, and 1:9. The three regions; I: below cac, II: above cac, but before saturation of the polymer, and III: above saturation of the polymer, are indicated in the figure.

Fluorescence measurements. The preparation of samples for fluorescence measurements was described earlier [13]. The pyrene concentration was kept low enough (<10^{-5} M) to prevent excimer formation. The DTBA concentrations were chosen to be less than one DTBA molecule per aggregate.

Dynamic fluorescence decay data were collected with the single photon counting technique with a set-up described elsewhere [14].

To determine the onset of surfactant aggregation, necessary for the calculations of the surfactant aggregation numbers, cac was determined for 0.1% PEO at various NaCl concentrations by surface tension measurements using the drop-weight method. For this purpose, an apparatus built at the department's workshop was used.

In order to check the drop-weight method, static fluorescence measurements were performed. It is shown [15] that the ratio between the third and the first vibronic peaks (denoted the III/I-ratio) in the pyrene steady-state fluorescence spectrum is dependent on the environmental polarity of the dissolved pyrene molecule. As aggregation starts, pyrene will be dissolved in the less polar aggregate and, thus, the III/I-ratio will change.

The method of dynamic fluorescence quenching in microheterogeneous solutions is well described in literature the [16,17]. Under the conditions that the excitation pulse is narrow compared to the fluorescence lifetime, the quenching in aggregates fast, and the aggregates small, spherical, and fairly monodisperse, the interpretation with the well-known Infelta-Tachiya model [18,19], or a generalized version [20,21], is straight-forward.

In the Infelta equation [18],

$$F(t) = B_1 exp[-B_2 t + B_3(exp(-B_4 t) - 1)]$$ (1)

the parameters have the following meaning with the assumptions stated above.

B_1 is the fluorescence intensity at time t=0, i.e., F(0).

B_2 is the decay rate at long time, i.e., when the decay shows an exponential tail. If there is no exchange of probe or quencher between the micellar phase and the bulk, B_2 will equal the unquenched decay rate k_0 [20].

B_3 equals the average number of quencher molecules per micelle if no exchange of probe or quencher occurs. If one knows the amount of bound surfactant molecules and the distribution of quenchers between micelles and the bulk phase, the aggregation number, <a>, can be calculated from

$$<a> = B_3 S_m/Q_m$$ (2)

where S_m is the bound surfactant concentration, i.e., the total surfactant concentration minus cac or cmc, and Q_m is the concentration of quencher in the aggregate sub-phase, i.e., the total quencher concentration minus the amount dissolved in the aqueous sub-phase.

B_4 equals the first-order quenching rate constant k_q if no exchange of probe or quencher occurs.

The unquenched lifetime, τ_0 (=1/k_0), was determined in separate experiments without added quencher, and the difference between 1/B_2 from quenching experiments and τ_0 is a measure of the condition that the probe and quencher are stationary during the time-window of the quenching experiments. This difference was close to zero, showing that no migration occured. Although the concentration of free quencher in the aqueous sub-phase thus was small, the fraction of quenchers which were free was significant at surfactant concentrations close to the cac, and had to be taken in account in calculation of the aggregastion numbers. To determine the distribution constant for the quencher DTBA between the aqueous and aggregate sub-phases, absorption measurements [22] were performed on different SDS-PEO solutions. Using the absorption in a saturated aqueous DTBA solution as background, the absorption from DTBA in the aggregates is given by

$$A_{agg} = A_{sample} - A_{aqueous} \tag{3}$$

The concentration of quencher associated with the aggregates, Q_m, was obtained using the absorptivity of DTBA in ethanol, determined as 105 $M^{-1}cm^{-1}$. The distribution constant K_D is given by

$$K_D = Q_m/(Q_{aq}S_m) \tag{4}$$

and with the value of the distribution constant known, all quencher concentration must be corrected for use in Eqn. 2 by

$$Q_m = Q_t S_m K_D/(S_m K_D+1) \tag{5}$$

where Q_t is the total quencher concentration.

The aggregation numbers obtained by Eqn. 2 should be treated as quencher-averaged aggregation numbers, $<a>_q$ [23-25]. The $<a>_q$ is an apparent property, dependent on Q_m. The weight-averaged aggregation number, $<a>_w$, independent of Q_m, is related to $<a>_q$ by [23,24]

$$<a>_q = <a>_w - 1/2\sigma^2\eta + 1/6\kappa\eta^2 - ... \tag{6}$$

where σ^2 is the variance and κ, the third cumulant, giving the skewness of the size distribution of the system. η is given by

$$\eta = Q_m/S_m \tag{7}$$

Eqn. 6 was used on the results from solutions with 100 mM and 500 mM NaCl and the reported aggretation numbers are $<a>_w$. The reported aggregation numbers for solutions with 50 mM added salt are the averages of the different $<a>_q$.

It should also be noted that the solutions for dynamic fluorescence measurements were not deoxygenated, which means that the fluorescence was quenched by bulk-water dissolved oxygen. This only affects the natural lifetime τ_0 and does not influence the use of Eqn. 1.

The computer programs used for the estimations of the natural lifetime and the parameters B_1-B_4 in Eqn. 1 are mainly based on the programs by Löfroth [26].

Dynamic light scattering (DLS) measurements have been made using the apparatus and techniques described in an earlier communication [27]. The data were assembled using an ALV wide-band, multi-tau, digital autocorrelator with 23 simultaneous sampling times allowing characterization of relaxation time distributions extending over 9 decades in time.

Data analysis. In the analysis of the measured autocorrelation curve, an inverse Laplace transformation (ILT analysis) was performed employing the algorithm REPES [28] to obtain the distribution of relaxation times. This program is similar to Provencher's CONTIN [29], except that the former directly minimizes the sum of the squared differences between experimental and calculated intensity-intensity $g_2(t)$ correlation functions using non-linear programming. For a system exhibiting a distribution of relaxation times, the field correlation function $g_1(t)$ $(g_2(t)=g_1^2(t)+1)$ is described by a continuous function of the relaxation time t using the Laplace transform:

$$g_1(t) = \int A(\tau)e^{-t/\tau}\, d(\tau) \tag{8}$$

Relaxation time distributions are given in the form of $\tau A(\tau)$ versus $\log \tau$ plots, with $\tau A(\tau)$ in arbitrary units.

The range of relaxation times allowed in the fitting was usually between 0.5 μs and 1 s with a density of 12 points per decade. Relaxation rates are obtained from the moments of the peaks in the relaxation time distribution or, if the peaks overlap, from the peak maximum position. With a broad distribution of relaxation times, these inversion methods yield multiple peaks in the "unsmoothed" analysis. The "smoothing" parameter (P) was selected as 0.5 in all cases, after it was established that the number of peaks did not increase with further increase in smoothing. As a further check, an analysis was made on a simulated correlation function consisting of a broad continuous distribution of relaxation times with noise added equal to the residuals from the analysis of the experimental correlation curve. REPES recovers the original distribution except when a very low smoothing parameter (P≈0) is used.

Diffusion coefficients are calculated as $(D = \Gamma/q^2)_{q->0}$ where q is the scattering vector $(q=[(4\pi n)/x]\sin(\Theta/2))$ and the relative amplitudes are obtained from the moments of the peaks and are given in the output of the program. Γ is the measured relaxation rate, $(\Gamma=\tau^{-1})$.

RESULTS and DISCUSSION

Critical aggregation concentration
From drop-weight measurements the cac in 0.1% PEO and the cmc in polymer-free solutions at different salt concentrations were calculated. The cac with no added salt was remarkably lower for the polymer used as compared to earlier findings [30], i.e., 5.5 mM for PEO with M_w in the range 12000-100000 and 2.9 mM for PEO with M_w = 960000.

The performed static fluorescence measurements given in Figure 2 as III/I-ratios, gave the same results, and the possibility that the difference was an effect of the drop-weight method was ruled out.

The cac-values obtained from both drop-weight and static fluorescence measurements are shown in Figure 3 and were used in the corrections for the aggregated surfactant concentrations.

The molecular weight dependence observed at the onset of aggregation cannot be explained by our investigations. It is an interesting question by itself, but need a more thorough treatment than the present.

DTBA distribution between the aqueous and aggregate sub-phases
The distribution constant K_D was determined by light absorption measurements to be

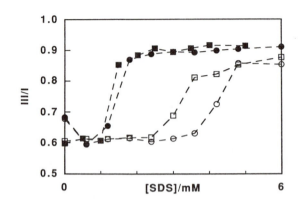

Figure 2. The ratio between the third and first vibronic peaks (III/I) in static fluorescence measurements. Circles refer to PEO with M_w=35000 and squares to PEO with M_w=600000. Open symbols refer to aqueous solutions without added salt, filled symbols to solutions with 100 mM NaCl.

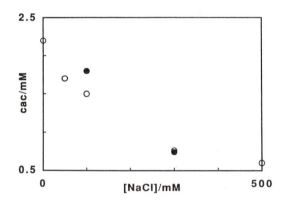

Figure 3. The critical aggregation concentration, cac, from static fluorescence and drop weight measurements versus salt concentration. Open circles refer to solutions with 0.1% PEO (M_w = 600000) and filled circles to solutions without PEO.

$1000 \, M^{-1}$, which means that at 1mM aggregated surfactant, 50% of the quenchers are solubilized in the aqueous phase; at 9 mM aggregated surfactant, 10% of the quenchers will be solubilized and at 99 mM aggregated surfactant, 1% of the quenchers will be solubilized in the aqueous phase. Thus, there is a substantial uncertainty implied in determining the aggregation numbers close to aggregation onset, an uncertainty which, due to the low value of the distribution constant, could be extended over a wide surfactant concentration range.

The distribution constant for DTBA between ordinary SDS micelles and water was determined to be $1800 \, M^{-1}$. The low value obtained when PEO is present, implies a substantial difference in the structure of the surfactant aggregate, even though the polymer does not penetrate the interior of the aggregate, but remains at the aggregate/water interface [9].

Pyrene fluorescence lifetime

The pyrene fluorescence lifetime, τ_0, Table 1 and Figure 4, shows a decrease with increasing SDS concentration. The decrease follows the same trend as was found in salt-free systems [12]. The reason for the shorter lifetime in this study as compared with the salt-free system (the τ_0 values from ref. 12 are inserted in Fig. 4 for comparison) is the difference in temperatures; 25 °C in this study and 20 °C in reference 12. It is interesting to note that the lower oxygen solubility in salt water [31] does not increase the pyrene fluorescence lifetime. Instead, it appears that the presence of NaCl causes the polymer to be less efficient in shielding the aggregate from bulk-water oxygen penetration. An important contribution to the lifetime behaviour in systems with added salt may be the fact that PEO forms complexes with alkali ions [32,33]. Thus, PEO will behave more like a polyelectrolyte in the presence of alkali salts, which would alter the interaction with the hydrophobic regions of the surfactant aggregate surface.

The longer lifetimes for the system with 500 mM salt seem to contradict this statement which may lead to the conclusion that the polymer interacts more strongly with the aggregate. The increase in lifetime is more likely to depend on a combination of lower oxygen solubility [31] and the fact that there is a change in the aggregate form at this high salt concentration; see below.

Aggregation numbers

Some typical DFQ-data are shown in Figure 5. The aggregate aggregation numbers, are shown together with results from references 10b and 12 in Figure 5 and summarized in Table 2. Note that the aggregation numbers presented in reference 12 have not been corrected for the solubility of the quencher. The correction, which has now been performed here , only results in slightly higher aggregation numbers at the lowest SDS concentrations, ≈ 20 and ≈ 30 for the uncorrected and corrected quencher distributions, respectively. In all comparisons with the results from reference 12, the corrected values are used.

The aggregation numbers at 0, 50 and 100 mM of added NaCl appears from Fig. 6 to be mainly determined by the SDS/PEO ratio. This may in part be a coincidence: all measurements with added salt except the one at SDS/PEO = 2 w/w were made close to or above the saturation concentration where the aggregation number approaches that of the free micelle (which increases somewhat with salt concentration). From Fig. 1 it can be found that the composition 0.1% PEO and 0.2% SDS corresponds to about one third of saturation, both with 100 mM salt and without added salt. It is more likely, therefore, that the degree of saturation controls the size rather than the SDS/PEO ratio. The results from Zana et al. [10b] inserted in Fig. 6, support the notion that the dependence on the SDS/PEO ratio is only apparent, depending on the

Table 1.
Pyrene fluorescence lifetimes obtained by dynamic fluorescence measurements in samples without added quencher. The results from salt-containing samples are from solutions with 0.1 % PEO at 25 °C, whereas the results from reference 12, the salt-free solutions, are from systems with 0.2 % PEO at 20 °C. In pure SDS micellar solutions, the lifetime is approximately 180 ns and 135 ns at 20 °C and 40 °C, respectively

[SDS] (%)	τ_0 (ns)		
	50 mM NaCl	100 mM NaCl	500 mM NaCl
0.20	193	187	203
0.40	185	184	197
0.50	182	181	
0.60	177	175	190
0.80	177	179	
1.00	178	170	185

0.2% PEO, no added salt (from reference 12)

0.26	217
0.50	210
0.58	205
0.72	203
1.73	192

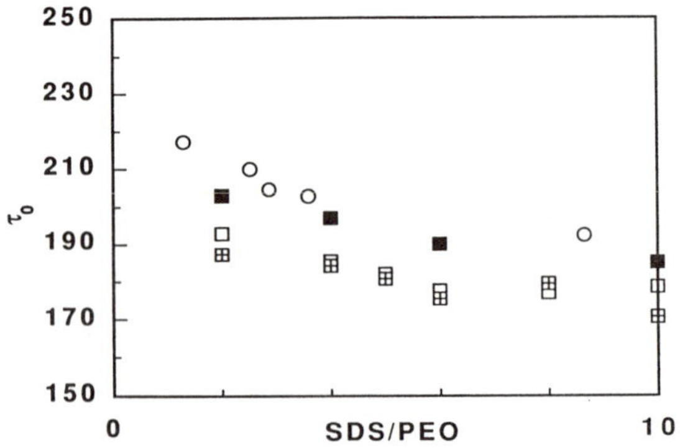

Figure 4. The natural lifetime, τ_0, versus SDS/PEO-ratio. Open circles: 0 mM NaCl added (from reference 12), open squares: 50 mM NaCl added, squares with crosses: 100 mM NaCl added, and filled squares: 500 mM NaCl added.

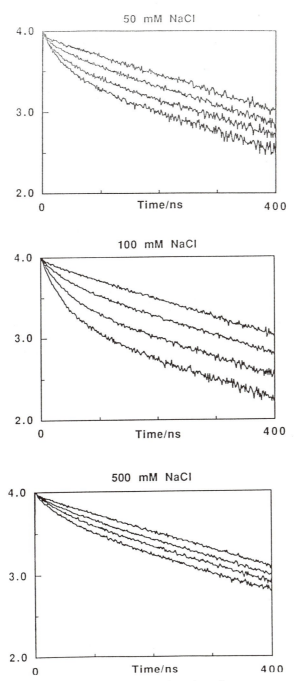

Figure 5. Dynamic fluorescence quenching data from systems with 0.1% PEO, 0.2% SDS, and 50, 100, and 500 mM added salt. The y-axis gives the logarithm of the normalized fluorescence intensity.

choice of concentrations; in the lower left corner of Fig. 6, the results from reference 10b deviate from the others. Fluorescence measurements at low SDS concentrations would be revealing in this respect, but are difficult to perform with reliability due to the water-solubility of the quencher.

The aggregation numbers at 500 mM NaCl exhibit a different pattern; they are markedly higher and increase over the whole SDS/PEO range. At this high salt concentration, SDS aggregates grow and change their form from small, spherical micelles to elongated prolates or hemi-sphere capped rods [34,35]. Evidently this is also the case when PEO is present, but with lower aggregation numbers than in polymer-free systems.

Quenching rate constants

The quenching rate constant, k_q, is thought to be, roughly, inversely proportional to the hydrophobic volume of the host aggregate, i.e., the aggregation number - this has been predicted by theory [36,37] and has been found experimentally valid, at least for systems with rather small, spherical, micelles [35], even if in some systems it has been found to depend on $V^{-1.2}$ [16].

Surprisingly, this is not valid in polymer-surfactant systems [12-14,38,39] – the quenching rate is rather independent of the aggregate aggregation numbers, Tables 2 and 3. Moreover, k_q is smaller in the surfactant-polymer systems than in the corresponding micellar systems, in spite of the fact that the aggregates have lower aggregation numbers than the micelles. One must remember when comparing the absolute values in Table 3, that the quencher used in reference 12 was not DTBA, but dimethylbenzophenone (DMBP). DMBP is a faster quencher than DTBA, and only trends in k_q can be compared.

These anomalies, found in several different surfactant-polymer systems under different conditions, clearly indicate that the interaction between the surfactant and the polymer slows down the dynamics of the solubilized molecules.

Zana et al. [10a] found that the pyrene excimer formation was similarly slowed down in SDS/PEO-aggregates. They proposed that an interaction between pyrene and PEO was the probable reason, and rejected an explanation based on increased "microviscosity" from the fact that the depolarization of diphenylhexatriene fluorescence was not affected.

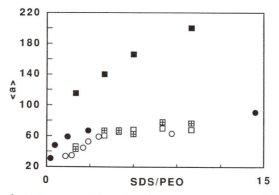

Figure 6. Surfactant aggregation numbers, $<a>_w$, versus SDS/PEO-ratio. Open circles: 0 mM NaCl added (from reference 12), filled circles: 0 mM NaCl added (from reference 10b), open squares: 50 mM NaCl added, squares with crosses: 100 mM NaCl added, and filled squares: 500 mM NaCl added.

Table 2.

Surfactant weight-average aggregation numbers for systems with 0.1% PEO and 50 mM, 100 mM and 500 mM NaCl from dynamic fluorescence quenching measurements. The aggregation numbers for 50 mM NaCl are the average of $<a>_q$, while the aggregation numbers for 100 mM and 500 mM salt are $<a>_w$, see text. Results from reference 12 are recalculated due to correction for the quencher distribution, see text. Ordinary SDS micelles have aggregation numbers around 55-65 in this concentration range in solutions with no added salt. With salt added larger aggregation numbers are obtained (up to 370 at =.6 M NaCl, reference 34)

[SDS]	$<a>$			$\sigma/<a>_w$	
%	50 mM	100 mM	500 mM	100 mM	500 mM
0.2	46.2	42.7	115	0.28	0.75
0.4	60.6	67.0	140	0.26	0.51
0.5	65.4	67.3		0.31	
0.6	68.2	62.5	166	0.38	0.41
0.8	69.9	78.0		0.35	
1.0	67.9	76.5	200	0.32	0.32

Salt-free systems, 0.2% PEO, from reference 12

[SDS]	$<a>_w$	$\sigma/<a>_w$
%		
0.26	33.8	0.76
0.35	34.9	0.35
0.50	44.8	0.37
0.58	52.8	0.37
0.72	59.2	0.44
1.73	63.3	0.40

Salt-free systems from reference 10b

[SDS]	0.2% PEO	[SDS]	0.5% PEO
%		%	
0.28	31	0.28	34
0.58	48	0.72	40
1.4	59	1.4	53
2.9	67	2.9	63
14	91	5.7	72
		9.6	79

[SDS]	1.0% PEO	[SDS]	2.0% PEO
%		%	
0.28	24	0.58	23
0.87	36	0.87	31
1.4	45	1.4	42
2.9	63	2.9	60
7.2	77	14	86
14	90		

The generality of the effect seems to rule out this proposal, however. It may well be that the depolarization measurement, probing the rotation of a molecule, is less affected than the translational diffusion involved in quenching and excimer formation.

Dynamic light scattering.

Diffusion coefficients were determined as $(\Gamma/q^2)_{q \to 0}$ after it had first been determined that the relaxation rate Γ was linearly dependent on the scattering vector q^2, passing through the origin, demonstrating a diffusive process.

Figure 7 shows typical data for the concentration dependence of D on c_{PEO}, with varying ratios of SDS and at a simple salt concentration of 100 mM. A constant ratio between the SDS and PEO concentration is used in order to approximate a constant binding degree at each PEO concentration.

The hydrodynamic virial coefficient, k_D, is defined as the concentration coefficient in: $D = D_0(1 + k_D c + ...)$; D_0 is the value of D at infinite dilution: $k_D = 2A_2M - k_f - v_2$ where M is the polymer molecular weight and k_f describes the concentration dependence of the friction coefficient, f where $f = f_0(1 + k_f c)$ and v_2 is the partial specific volume. k_D is thus the sum of a static factor, proportional to the second virial coefficient, A_2, and the concentration dependence of the friction coefficient. Expressions for k_f have been given and are summarized in [40]. For a model of interacting hard spheres, the pair interaction potential describing the virial coefficient is usually given by the DLVO

Table 3.

Quenching rate constants from dynamic fluorescence quenching measurements with di-tert-butylaniline as quencher and pyrene as fluorescence probe.
The quenching rates from reference 12 are for dimethylbenzophenone as quencher

[SDS]	$k_q \cdot 10^{-7}$ s^{-1}		
%	50 mM	100 mM	500 mM
0.2	2.0	2.3	1.5
0.4	1.7	1.9	1.2
0.5	1.7	1.9	
0.6	1.7	1.9	0.9
0.8	1.7	1.8	
1.0	1.7	1.8	0.6

[SDS]	Salt-free systems, 0.2% PEO, from reference 12 $k_q \cdot 10^{-7}$ s^{-1}
%	
0.26	4.5
0.35	4.3
0.50	3.7
0.58	3.6
0.72	3.3
1.73	3.4

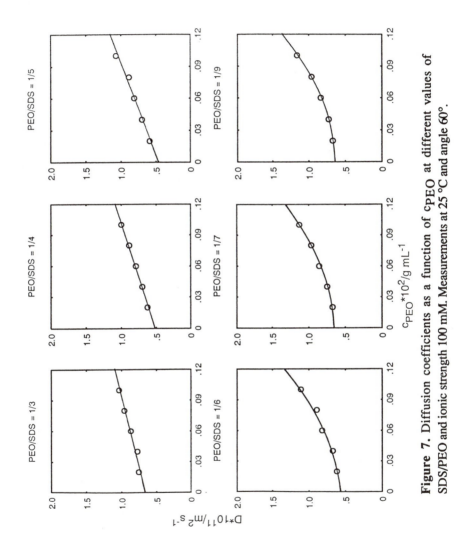

Figure 7. Diffusion coefficients as a function of c_{PEO} at different values of SDS/PEO and ionic strength 100 mM. Measurements at 25 °C and angle 60°.

theory as the sum of a repulsive interaction due to the charges on the spheres and an attractive interaction (van der Waals forces). Approximate expressions are obtained by solution of the Poisson-Boltzmann equation in limiting cases. Thus k_D initially increases as shown in Figure 8a as the repulsive contributions to the interaction potential due to the fixed charges become increasingly important and subsequently decreases owing to the screening effect of free sodium counterions from the SDS which progressively increase in concentration. Figure 8b shows the corresponding variation of the second virial coefficient A_2 from intensity light scattering with the SDS/PEO-ratio. The trend in A_2 follows closely that in k_D in Fig. 8a.

Hydrodynamic radii for each SDS/PEO-value obtained at the limit of $c_{PEO}=0$ from the data in Fig. 6 at 100 mM NaCl according to the Stokes-Einstein equation: $R_H=kT/6\pi\eta_0 D_0$ are depicted in Figure 9 where R_H is plotted versus SDS/PEO. The location of the maximum in the hydrodynamic radius represents the point at which the binding reaches saturation. The hydrodynamic radius for the saturated complex at 100 mM salt, $R_H =550$ Å, can be compared with an extrapolated value of the radius of gyration from Cabane and Duplessix [9b]; according to which $R_G \approx 700$ Å should apply for a PEO with a molecular weight of 10^6 g/mole saturated with SDS in 400 mM salt. The chain at the lower salt concentration would be expected to be somewhat more extended, and the true R_H/R_G ratio for the low salt concentration somewhat smaller than the value of 0.78, which would be appropriate for a flexible polymer in a good solvent.

Cabane and Duplessix also reported a second virial coefficient of $6 \cdot 10^{-5}$ cm^2 mole g^{-2} for the saturated chain in 400 mM salt solution. This value is much smaller than those in Fig. 7b for 100 mM salt solutions, which is expected due to the greater screening of the electrostatic repulsion.

Figure 10 shows the influence of low molar mass salt on the diffusion coefficient where the data are displayed as a function of c_{PEO} showing how the interaction potential decreases with ionic strength. D decreases with increasing ionic strength due to the screening effect.

Figure 11 shows the accumulated data from plots of the type of Fig. 10 as a function of c_{SDS}. The overall form of the curves shown in Fig. 11 is dependent on the screening effect of free ions on the interactions between the fixed charges, i.e., it is simply related to the pair interaction potential. D initially increases with c_{SDS} as the degree of binding and hence the interaction potential increases up to the point of saturation. Subsequently, D decreases at a given ionic strength owing to screening by free counterions. At the lowest ionic strength (0.3 mM), the maximum value of D is attained at SDS/PEO≈4 w/w, which value agrees approximately with that of reference 12 - Fig. 6 - in salt-free solutions. Screening by the low molecular weight salt leads to a lower D-value for a given SDS concentration and the maximum moves as anticipated to higher SDS concentration.

The scattering experiments were mainly performed using constant SDS/PEO-ratios, and diluting the solutions at constant salt concentration from 0.1% PEO. In general the degree of saturation of the polymer changes somewhat in this procedure. From the dilution lines indicated in Fig. 1 it is apparent that the 1:5 ratio is close to the suggested saturation limit att 100 mM NaCl. It is at this composition that most of the results, i.e., k_D, A_2, and R_H, have their peak values, supporting saturation close to this ratio. The progressive displacement of the peak of the diffusion coefficient in Fig. 11 with salt concentration is also in general agreement with the expected change of saturation limit.

Relaxation time distributions obtained using inverse Laplace transformation (ILT) with the algorithm REPES (Experimental) are shown in Figure 12 for data over a range of SDS/PEO up to 20. As shown in this and the subsequent diagrams, peaks

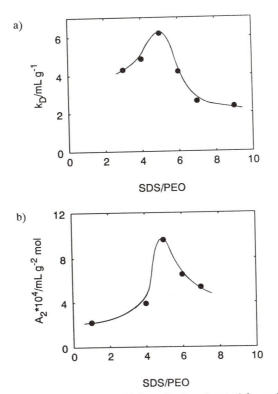

Figure 8a. Hydrodynamic virial coefficient (k_D) estimated from the initial slopes in Fig. 7 as a function of SDS/PEO. **b.** Second virial coefficients (A_2) obtained from intensity light scattering as a function of SDS/PEO.

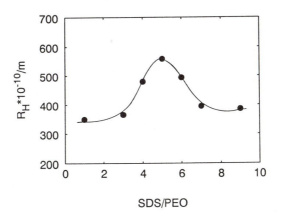

Figure 9. Hydrodynamic radius as a function of SDS/PEO; 100 mM NaCl and 25 °C.

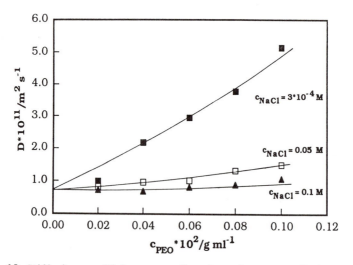

Figure 10. Diffusion coefficients as a function of c_{PEO} at SDS/PEO = 1:5 at different ionic strengths. Data at 25 °C and angle 60°.

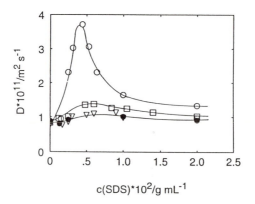

Figure 11. Diffusion coefficient for the SDS/PEO-complex as a function of c_{SDS} at a constant concentration c_{PEO}=0.1%. Data at different ionic strengths as shown, 25 °C and at angle 60°

are registered for both the SDS/PEO-complex and free SDS micelles. Thus the lower intensity peak at shorter relaxation time represents free SDS micelles and the main peak the SDS/PEO-complex. The estimated hydrodynamic radius as indicated for the SDS peak agrees well with the value obtained for SDS alone in aqueous solution at that ionic strength; see below.

The appearance of a free micelle peak in the distributions shown in Figures 12-14 give further support to the phase diagram. The results in Fig. 12 all pertain to compositions in region III were free micelles should be abundant. Fig. 13a, with SDS/PEO 1:3 w/w, passes from region II, with unsaturated polymer chains, into region I at the two lowest PEO concentrations, i.e., below the cac. These two curves peak at the same relaxation time, which should be similar to that for the polymer itself.

At 1:9 SDS/PEO, Fig. 13b, the lowest PEO concentration pertains to a composition within region II, close to the cac. There should not be any free micelles present, accordingly, and the appearance of a micelle peak there may be an indication of the uncertainty of the phase diagram at low polymer concentrations.

The effect of the change of the salt concentration on the relaxation time distribution at SDS/PEO 1:9 w/w, shown in Fig. 14, appears larger than may be expected from the change in the phase diagram. At both salt concentrations this composition is far beyond the saturation limit. The amplitudes of the scattering peaks should, however, directly reflect the relative amounts of material present as free micelles in the polymer-surfactant complex when the main peak areas are equal.

Aqueous SDS solutions.

Complementary measurements were made on SDS in NaCl solutions. Figure 15a shows D as a function of SDS concentration at three salt concentrations. There is an apparent increase in the hydrodynamic radius as the salt concentration is raised, from a value of 16 Å in 0.3 mM NaCl to 27 Å in 100 mM NaCl. As shown by fluorescence probing, this increase is not due to a larger aggregation number at the higher ionic strength, but to an increase in the diffusion coefficient due to the electrostatic interaction between the micelles at the low ionic strength [41]. These data are in good agreement with those of Maser et al. [34]. Typical relaxation time distributions are depicted in Figure 15b for data at different SDS concentrations in the range up to 5% in 10 mM NaCl. The distributions were always single-peaked and no secondary aggregates are present.

Conclusions

The critical aggregation concentration changes to a lower value if the polymer molecular weight is sufficiently high.

The aggregates formed on the cooperative binding of SDS to PEO are initially small and grow to a size slightly smaller than that of free micelles on the saturated chain. With increasing salt concentration, the amount of SDS bound on saturation is greatly increased; the size of the aggregates remains small, however, which implies that more aggregates form on each chain. At the highest salt concentration, 500 mM, SDS form elongated prolate particles or rods with higher aggregation numbers. The change in aggregate structure also changes the lifetime of the fluorescence probe, pyrene, to higher values.

The aggregates formed on the chain do not solubilize small molecules (2,4-di-tert-butylaniline and dimethylbenzophenone) as well as the free micelles, and the translational diffusion of the guest molecules, required for quenching, is greatly reduced, suggesting a more rigid packing than in the free micelles.

The SDS/PEO-complex behaves as a flexible chain in a good solvent, with maximum expansion at saturation; the expansion is strongly reduced with increasing ionic strength up to above 100 mM.

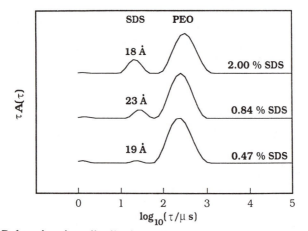

Figure 12. Relaxation time distributions obtained by ILT for the SDS/PEO-complex as a function of SDS concentration at ionic strength 50 mM. Constant PEO concentration c_{PEO}=0.1% and 25 °C.

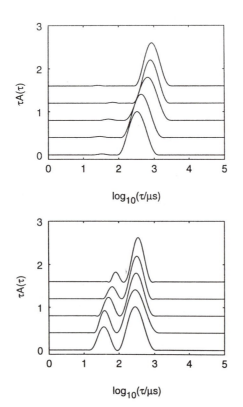

Figure 13. Relaxation time distributions using ILT for different PEO concentrations at 10 mM NaCl and 25 °C:
a. at SDS/PEO = 1:3; **b.** at SDS/PEO =1:9.

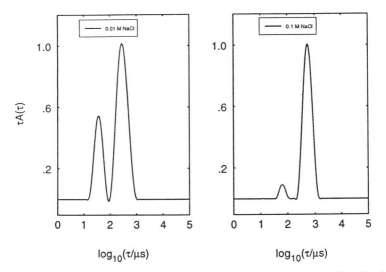

Figure 14. Influence of NaCl concentration: relaxation time distributions at SDS/PEO=1:9 with c_{PEO}=0.1%. Data at 25 °C.

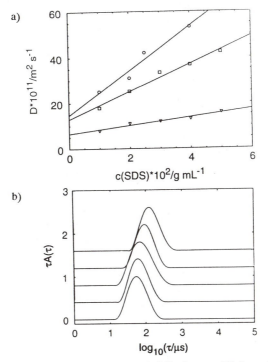

Figure 15a. Concentration dependence of diffusion coefficients for SDS micelles in media of different ionic strength at 25 °C. **b.** Relaxation time distributions using ILT for SDS micelles at different concentrations in 10 mM NaCl at 25 °C.

ACKNOWLEDGEMENTS

Grants from The Swedish Natural Science Research Council (NFR) and The Swedish National Board for Industrial and Technical Development (NUTEK) were used to finance this project. We are grateful for the help of Ms. Anna Lindahl and Ms. Charlotta Naeslund in performing the absorption measurements when determining the quencher distribution constants and to Mr. Göran Svensk, Ms. Helena Berglund and Ms. Anna Sjöström with performing the drop-weight measurements when determining the critical aggregation concentrations. We thank Mr. Göran Karlsson for help with the figures. We are thankful for the fruitful discussions with and help from Professors Hans Vink, Bertil Enoksson, and Gerd Olofsson.

References

1. Goddard E.D., *Coll. Surf.*, 19, 255 (1986)
2. Robb I.D., *Anionic surfactants in physical chemistry of surfactant action*, Lucassen-Reynders E. (Ed), Marcel Dekker, New York, pp. 109 (1981)
3. Saito S., *Nonionic Surfactant, Physical Chemistry Surfactant Science Series*, Schick M.J. (Ed), Marcel Dekker, New York, pp. 991 (1987)
4. Lindman B., Karlström G., *The Structure, Dynamics and Equilibrium Properties of Colloidal Systems*, Bloor D.M., Wyn-Jones E. (Eds), Kluwer Academic Publishers, Doordrecht, pp. 131 (1990)
5. Lindman B., Thalberg K., *Polymer-Surfactant Interactions*, Goddard E.D., Ananthapadmanabhan K.P (Eds), CRC Press, Boca Raton, Florida, in press
6. Piculell L., Lindman B., *Adv. Coll. Pol. Sci.*, submitted
7. Jones M.N., *J. Coll. Int. Sci.*, 23, 36 (1967)
8. Schwuger M.J., *J. Coll. Int. Sci.*, 43, 491 (1972)
9. a. Cabane B., *J. Phys. Chem.*, 81, 1639 (1977)
   b. Cabane B., Duplessix R., *J. Phys.*, 43, 1529 (1982)
   c. Cabane B., Duplessix R., *Coll. and Surf.*, 13, 19 (1985)
   d. Cabane B., Duplessix R., *J. Phys.*, 48, 651 (1987)
10. a. Zana R., Lianos P., Lang J., *J. Phys. Chem.*, 89, 41 (1985)
    b. Zana R., Lang J., Lianos P., *Microdomains in polymer solutions*, Dubin P. (Ed) Plenum Press, New York and London, pp.357 (1985)
11. Winnik F.M., Winnik M.A., *Pol. J*, 22, 482 (1990)
12. van Stam J., Almgren M., Lindblad C., *Progress in Colloid and Interface Science*, 84, 13 (1991)
13. Thalberg K., van Stam J., Lindblad C., Almgren M., Lindman B., *J. Phys. Chem.*, 95, 8977 (1991)
14. Almgren M., Hansson P., Mukhtar E., van Stam J., *Langmuir*, submitted
15. Kalyanasundaram K., Thomas J.K., *J. AM. Chem. Soc.*, 101, 2039 (1977)
16. Almgren M., *Kinetics and catalysis in Microheterogenous Systems*, Grätzel M., Kalyanasundaram M. (Eds), Marcel Dekker, New York, pp. 63 (1991)
17. Zana R., *Surfactants in Solution, vol 22*, Zana R. (Ed), Marcel Dekker, New York and Basel, pp 241 (1987)
18. a. Infelta P.P., Grätzel M., Thomas J.K., *J. Phys. Chem.*, 78, 190 (1974)
    b. Infelta P.P., Grätzel M., *J. Phys. Chem.*, 78, 5280 (1983)
19. a. Tachiya M., *Chem. Phys. Lett.*, 33, 289 (1975)
    b. Tachiya M., *J. Chem. Phys.*, 76, 340 (1982)
    c. Tachiya M., *J. Chem. Phys.*, 78, 5282 (1983)
20. Almgren M., Löfroth J.-E., van Stam J., *J. Phys. Chem.*, 90, 4431 (1986)
21. Gehlen M.H., Van der Auweraer M., De Schryver F.C., *Langmuir*, 8, 64 (1992)
22. Almgren M., Alsins J., *Progr. Colloid Pol. Sci.*, 74, 55 (1987)
23. Almgren M., Löfroth J.-E., *J. Chem. Phys.*, 76, 2734 (1982)
24. Warr G.G., Grieser F., *J. Chem. Soc., Faraday Trans. 1*, 82, 1813 (1986)

25. a Almgren M., Alsins J., Mukhtar E., van Stam J., *J. Phys. Chem.*, 92, 4479 (1988)
    b Almgren M., Alsins J., Mukhtar E., van Stam J., *Reactions in Compartmentalized Liquids*, Knoche W., Schomäcker R. (Eds), Springer-Verlag, Berlin & Heidelberg, pp 61 (1989)
26. Löfroth J.-E., *Eur. Biophys. J.*, 13, 45 (1985)
27. Brown W., Schillén K., Almgren M., Hvidt S., Bahadur P., *J. Phys. Chem.*, 95, 1850 (1991)
28. a Jakes J., *Czech J. Phys.*, B38, 1305 (1988)
    b Johnsen R. M., Brown W., *Laser Light Scattering in Biochemistry*, Harding S. E., Sattelle D. B., Bloomfield V. A., (Eds), Royal Society of Chemistry, pp.77 (1992)
29. Provencher S. W., *Makromol. Chem.*, 180, 201 (1979)
30. Lindblad C., van Stam J., unpublished results
31. Landolt-Börnstein, 6th ed, Springer Verlag, Berlin Göttingen Heidelberg, 1962
32. Poonia N.S., Bajaj A.V., *Chem. Rev.*, 79, 389, 1979
33. Dubin P.L., Gruber J.H, Xia J., Zhang H., *J. Coll. Int. Sci.*, 148, 35 (1992)
34. Mazer N. A., Benedek G. B., Carey M. C., *J. Phys. Chem.*, 80, 1075 (1976)
35. Almgren M., Löfroth J.-E., *J. Coll. Int. Sci.*, 81, 486 (1981)
36. Dederen J., van der Auweraer M., De Schryver F.C., *Chem. Phys. Lett.*, 68, 451 (1979)
37 van der Auweraer M., Dederen J., Geladé E., De Schryver F.C., *J. Chem. Phys.*, 74, 1140 (1981)
38. Zana R., Lianos P., Lang J., *J. Phys. Chem.*, 89, 41 (1985)
39. van Stam J., Almgren M., Lindblad C., Mukhtar E., Adolfsson A., Martins Miguel M., in preparation
40. Felderhof B.U., *J. Phys. A: Math. Gen.*, 11, 929 (1978)
41. Pusey P. N., *Photon Correlation and Light Beaming Spectroscopy*, Cummins H. Z., Pike E. R. (Eds), Plenum Press, New York, pp. 58 (1956)

RECEIVED May 24, 1993

Chapter 17

Interactions of Hydrophobically Modified Poly(N-Isopropylacrylamides) with Liposomes

Fluorescence Studies

H. Ringsdorf[1], J. Simon[1], and F. M. Winnik[2,3]

[1]Institut für Organische Chemie, Johannes Gutenberg-Universität Mainz, J.-J. Becher Weg 18–20, D–6500 Mainz, Federal Republic of Germany
[2]Xerox Research Center of Canada, 2660 Speakman Drive, Mississauga, Ontario L5K 2L1, Canada

The interactions of Dimyristoylphosphatidylcholine (DMPC) and distearoylphosphatidylcholine (DSPC) small unilamellar vesicles and hydrophobically modified poly-(N-isopropylacrylamides) have been examined in water at 25°C with mixtures of random copolymers of N-isopropylacrylamide and N-[4-(1-pyrenyl)butyl]-N-n-octadecylacrylamide (PNIPAM-C_{18}Py/200, and NIPAM and N-[2-(1-naphthyl)ethyl]-N-n-octadecyl-acrylamide (PNIPAM - C_{18}Na/200). These polymers form interpolymeric micelles in water. From fluorescence experiments using non-radiative energy transfer between excited naphthalene and pyrene, it was established that the polymeric micelles are disrupted irreversibly upon contact with the liposomes. The polymer chains reorganize on the surface of the liposome in a slow process which results in the spatial separation of the octadecyl groups previously held in close contact within the polymeric micelles. Upon saturation coverage of the liposome surface, coated liposomes and free polymeric micelles coexist in solution. Analogous experiments with labeled derivatives of the NIPAM homopolymer reveal that a different mechanism is operative with polymers that do not carry octadecyl substituents.

Liposomes form spontaneously when lipids are dispersed in aqueous media. They consist of a bilayer membrane which captures a pool of water in its center. The practical value of liposomes derives from two unique properties: 1) their ability to entrap either water-soluble materials in the internal water reservoir or liposoluble compounds in the lipid bilayer; and 2) their compatibility with natural membranes, making them safe for medical application. They have attracted considerable interest as delivery systems, such as carriers of drugs to specific targets in the treatment of cancer tumors and in enzyme therapy (1) and in various diagnostic tests using fluorescent makers (2) and radiolabels (3). Liposomes are fragile structures created in water as a result of a delicate balance of interacting forces which arise when amphiphilic compounds are added to water. In most practical applications liposomes have to be

[3]Corresponding author

0097–6156/93/0532–0216$07.25/0

stabilized so that they conserve their integrity in hostile environments. They must also respond to an external stimulus and deliver the entrapped materials once they have reached their target.

Clearly, key to the success of the liposome technology are systems with improved mechanical stability in which the dynamical properties of the membrane are preserved and can be controlled.   One option explored in several laboratories consists in adsorbing securely on the outer surface of the liposome a polymer which responds by some physical change to an external stimulus, such as a pulse of light or a change in pH or   conductivity of the suspending medium (4).   The polymer layer often imparts structural stability to the liposome (5).   We have examined the use of thermoresponsive polymers in such systems (6).   The   structure   of   the   polymers   consists   of   a   poly-(N-isopropylacrylamide) chain carrying at random a few pendent octadecyl groups.   The   alkyl   substituents   serve   as   anchors,   the   N-isopropylacrylamide   chain   provides   the   thermosensitivity.   Poly-(N-isopropylacrylamide)   (PNIPAM)   exhibits   a   lower   critical   solution temperature (LCST) in water at 31°C.   Detailed investigations of this phase transition established that at the LCST the PNIPAM chains undergo a collapse from hydrated extended coils to hydrophobic globules which aggregate and form a separate phase (7).   This contraction/expansion can be exploited in liposome systems in the following way: 1) the heat-induced collapse of the chains anchored on the surface of a liposome triggers the contraction of the external lipid bilayer and 2) the constraints thus imposed are relieved as the polymer chain re-expands upon cooling.   This active role of anchored NIPAM copolymers was ascertained by monitoring the thermally-induced changes in the photophysical properties of a fluorescently labeled amphiphilic NIPAM copolymer.

While these initial experiments provided good evidence in support of this "molecular accordion" mechanism, many issues remained to be clarified.   Here we examine the mechanism of the adsorption and docking of the polymers as they are added, below the LCST, to aqueous liposome suspensions. Special attention was paid to the role of the alkyl pendent groups during the adsorption process and to the importance of the physical state of the lipid bilayer.   These questions were addressed primarily through the use of fluorescence measurements with labeled polymers.   Supportive evidence was provided by microcalorimetric studies.

Experimental

Materials.   Water was deionized with a Millipore Milli-Q water purification   system   (specific   conductance: 0.056 μmohs   cm-1). Dimyristoylphosphatidylcholine   (DMPC)   and distearoylphosphatidylcholine   (DSPC)   were   purchased   from   Sigma Chemicals.   HPLC grade chloroform (Caledon) was used in the liposome preparation. The synthesis and characterization of the labeled polymers were   described   in   the   following   publications: PNIPAM-Py/200   (8), PNIPAM-Na/37 (9),   PNIPAM-C_{18}-Py/200 (10), PNIPAM-C_{18}-Na/200 (11), PNIPAM-C_{18}-Na/680-C_{18}Py/1700 (11).

Fluorescence Measurements.   Steady-state   fluorescence   spectra were recorded on a SPEX Fluorolog 212 spectrometer equipped with a DM3000F data system.   The temperature of the water-jacketted cell holder was controlled with a Neslab circulating bath. The temperature of the sample fluid was measured with a thermocouple immersed in the sample. Emission spectra were recorded with an excitation wavelength of 330 nm (pyrene) and 290 nm (naphthalene) and were not corrected. The pyrene excimer to monomer ratios I_E/I_M were calculated by taking the ratio of the intensity (peak height) at 480 nm to the half-sum of the intensities at 379 nm and 399 nm. The monomer emission was determined in each case by subtracting a normalized (at 343 nm) spectrum. The extent of pyrene emission due to non-radiative energy transfer (NRET) from naphthalene is reported here in terms of the ratio I_{Na}/I_{Py} of the intensity (peak height) at 340 nm to the intensity at 378 nm of the emission spectra from excitation at 290 nm.

Preparation of the liposomes. Single unilamellar vesicles (SUV) were prepared by the sonication technique (12). A solution of the lipid (DSPC or DMPC, 10 mg) in chloroform (10 mL) was evaporated by the use of a rotary evaporator to form a thin dry film on the wall of a tube. Deionized water (10 mL) was added to disperse the film by gentle shaking. The mixture was heated above the lipid phase transition temperature for 30 min (heating temperature: 25°C for DMPC and 70°C for DSPC). The cooled suspension was subjected to vortex mixing for 1 minute. It was disrupted by sonication for 4 minutes with a Cole Parmer Ultrasonic Homogenizer equipped with a standard tip Series 4710. Sonication was performed at a power of 35 W. The clear suspension was allowed to stand for 2 hours above the phospholipid transition temperature. After cooling to room temperature, the solution was filtered through a 0.22 μm filter. Samples were used only if they retained a constant size for a period of at least 6 h prior to the spectroscopic measurements.

Samples for Spectroscopic Analysis. Different procedures were used. In all cases polymer powders were allowed to dissolve in water for 1h at room temperature, then the solutions were kept at 5°C for 24 hr to ensure complete dissolution of the polymers. In experiments with hydrophobically-modified NIPAM copolymers, solutions were prepared from aqueous stock solutions (1 g L^{-1}) diluted in deionized water in the desired ratios and amounts. Solutions of PNIPAM-Py/200 and PNIPAM-Na/40 of lower concentration (0.1g L^{-1}) were prepared and used without further dilution. The polymer solutions were added to a (filtered) liposome suspension in deionized water. The resulting suspensions were allowed to stand at room temperature for 3 hours, unless otherwise indicated.

Microcalorimetric Measurements. Differential scanning calorimetry was performed with a Microcal MC 2 DSC microcalorimeter. The samples, prepared as described above, were analyzed using a heating rate of .0.75°C min^{-1} in the temperature range of 10 to 40°C for DMPC liposomes and 25 to 65°C for DSPC liposomes.
Liposome Size Measurements. Sizes were determined at 25°C by quasi-elastic light scattering (QELS) with a fixed 90° scattering angle using a Brookhaven Instrument Corporation Particle Sizer Model BI-90 equipped with a He/Ne laser. In a typical measurement values were determined over 2000 cycles with a count rate of 50 kpcs. The software provided by

the manufacturer was employed to determine the liposome sizes. An average value over 10 consecutive measurements is reported.

Materials and Spectroscopy.

Synthesis and Structure of the Polymers. Three amphiphilic NIPAM copolymers were used in this study (Figure 1). They were prepared by free-radical copolymerization in dioxane of NIPAM and the corresponding chromophore-substituted acrylamides. They are PNIPAM-$C_{18}Py/200$, a random copolymer of N-isopropylacrylamide (NIPAM) and N-[4-(1-pyrenyl)butyl]-N-n-octadecylacrylamide; PNIPAM-$C_{18}Na/200$, a random copolymer of N-isopropylacrylamide and N-[2-(1-naphthyl)ethyl]-N-n-octadecylacrylamide; and a doubly labeled copolymer PNIPAM-$C_{18}Na/680$-$C_{18}Py/1700$, a random copolymer of N-isopropylacrylamide, N-[4-(1-pyrenyl)butyl]-N-n-octadecylacrylamide, and N-[2-(1-naphthyl)ethyl]-N-n-octadecyl-acrylamide. The molar ratio of chromophores to NIPAM units is small in all cases, for example 1:200 for PNIPAM-$C_{18}Py/200$. Physical properties and compositions of the labeled polymers are listed in Table I.

The homopolymer, poly-(N-isopropylacrylamide) and its labeled derivatives were needed to carry out control experiments. The homopolymer was prepared by free-radical polymerization of NIPAM in dioxane or tert-butanol. Attachment of the dyes to the PNIPAM backbone was achieved by reacting a copolymer of N-isopropylacrylamide and N-acryloxysuccinimide with chromophores bearing a short amino-terminated alkyl chain. Specifically, the polymer was labeled with pyrene by reaction with 4-(1'-pyrenyl)-butylamine or with naphthalene by reaction with 1-(1'-naphthyl)-ethylamine. The molecular weights of PNIPAM and its labeled derivatives were significantly larger than those of the amphiphilic copolymers (Table I).

Spectroscopy of the Polymers in Solution. The emissions of the pyrene and naphthalene groups attached to the polymers are sensitive to small changes in the chromophore separation distances. A short separation distance (ca 4 to 5 Å) can be monitored with pyrene labeled polymers via changes in the features of the pyrene emission, and a longer scale (ca 15 to 50 Å) by measuring the extent of non-radiative energy transfer between the two chromophores, either in solutions of the doubly-labeled copolymer or in mixed solutions of pyrene- and naphthalene-labeled materials.

Pyrene Monomer and Excimer Emission. The emission of locally isolated excited pyrenes ('monomer' emission, intensity I_M) is characterized by a well-resolved spectrum with the [0, 0] band at 378 nm. The emission of pyrene excimers (intensity I_E), centered at 480 nm, is broad and featureless. Excimer formation requires that an excited pyrene (Py*) and a pyrene in its ground state come into close proximity within the Py* lifetime. The process is predominant in concentrated pyrene solutions or under circumstances where microdomains of high local pyrene concentration form, even though the total pyrene concentration is very low. This effect is shown for example by comparing the spectra of the hydrophobically modified, pyrene labeled polymer, PNIPAM-C_{18}-Py/200, in water and in methanol. The strong excimer emission from the aqueous solution ($I_E/I_M = 1.10$) vouches for

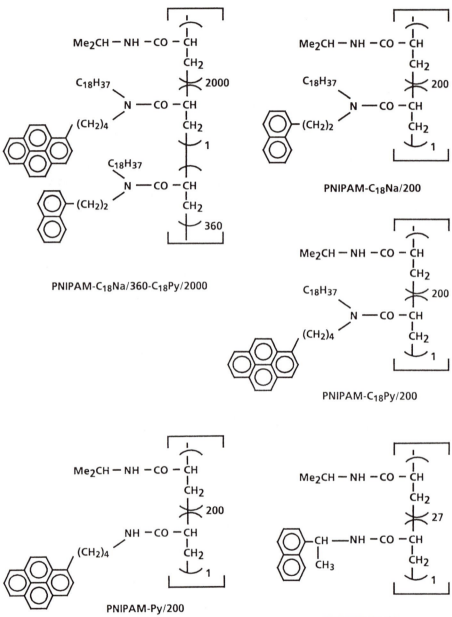

Figure 1:   Chemical structures of the polymers used in this study.

Table I. Molecular and Physical Properties of the Polymers

Polymer	Composition (unit mol) NIPAM: C_{18}Na:C_{18}Py	[Na] mol g^{-1} mol	[Py] mol g^{-1}	M_v	LCST (°C)	Ref.
PNIPAM				400,000	31.8	22
PNIPAM-C_{18}Na/200	230 (205) : 1 : ---	4.0×10^{-5}		279,000	30.4	22
PNIPAM-C_{18}Py/200	206 (181) : -- : 1		4.4×10^{-5}	390,000	30.6	20
PNIPAM-C_{18}Na/360-C_{18}Py/2000	2040 340 : 5.6 : 1 (C_{18}) : 1	2.4×10^{-5}	0.4×10^{-5}	335,000	30.8	22
PNIPAM-Na/27	(27) : 1 : --	3.26×10^{-4}		1.2×10^6	32.5	9
PNIPAM-Py/200	(175) : -- : 1		4.2×10^{-5}	1.1×10^6	33.0	8

the presence of hydrophobic microdomains in which the pyrene groups come in close contact. By contrast the emission of pyrene in methanolic solution of PNIPAM-C_{18}-Py/200 displays mostly pyrene monomer emission (I_E/I_M = 0.04), the more common occurrence in dilute solutions. Hence disruption of the microdomains, by surfactants or co-solvents for example, resulting in a local dilution of the chromophores can be detected readily by a large decrease in excimer emission concomitant with an increase in pyrene monomer emission (Figure 2).

The fluorescence spectra of the pyrene labeled polymer which does not carry pendent octadecyl chains, PNIPAM-Py/200, are presented in Figure 3 for solutions in water and in methanol. In methanol, the emission originates mostly from locally isolated excited pyrenes (I_E/I_M = 0.06). In water the relative contribution from the excimer is much larger (I_E/I_M = 0.40). Although not apparent from Figure 3, the overall emission intensity is lower in water, indicating a large extent of pyrene self-quenching. The excimer emission in this case originates mostly from pre-formed pyrene dimers and not via dynamic encounter of Py* and Py, in accordance with the classic mechanism described by Birks (13) and exemplified by the pyrene emission from methanolic solutions of PNIPAM-Py/200 and PNIPAM-C_{18}-Py/200 and from aqueous solutions of PNIPAM-C_{18}Py/200 in water. Pyrene dimers have been detected in aqueous solutions of several pyrene-labeled polymers (14). Their existence has been attributed to a gain in free energy of mixing through hydrophobic interactions between the non-polar pyrene groups. Their occurrence is revealed by the following spectral features. The excitation spectra for the monomer and the excimer emissions are different. Their general features are similar, but the former is blue-shifted by about 3 nm (Figure 3, top inset). Furthermore the time-dependent excimer fluorescence profile does not show a growing-in component, at least in the nanosecond time-scale. In solutions of PNIPAM-Py/200 pyrene-pyrene aggregates occur predominantly among groups attached to the same polymer chain.

Non-Radiative Energy Transfer (15, 16). This process originates in dipole-dipole interactions between an energy donor in its excited state and an energy acceptor in its ground state. The probability of energy transfer between two chromophores depends sensitively on their separation distance and to a lesser extent on their relative orientation. Therefore in mixed solutions of polymers carrying either donors or acceptors the extent of energy transfer between the two labels can be related to the extent of interpolymeric association. In solutions of polymers carrying both chromophores along the same chain, variations in NRET provide information on changes in the conformation of the chain, such as coil collapse or expansion. The application of NRET to probe interpolymeric interactions in solutions of "man-made" polymers was pioneered by Nagata and Morawetz (17). Indeed the technique has been employed extensively in biochemistry, for example in studies of the tertiary structure of proteins (18), or to monitor the fusion of bilayer vesicles (19). It has been powerful also in the present study. Its application to probe mixed solutions is reviewed briefly here in the case of the hydrophobically-modified PNIPAM-C_{18}-Na/200 and PNIPAM-C_{18}-Py/200. The pyrene-naphthalene pair of chromophores is known to interact as energy donor (naphthalene) and energy acceptor (pyrene) by non-radiative energy transfer with a characteristic distance, R_0 = 29 Å, R_0

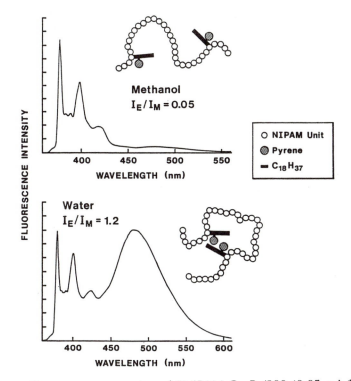

Figure 2:   Fluorescence spectra of PNIPAM-C_{18}Py/200 (0.05 g L^{-1}) in water and in methanol; λ_{exc} = 330 nm.

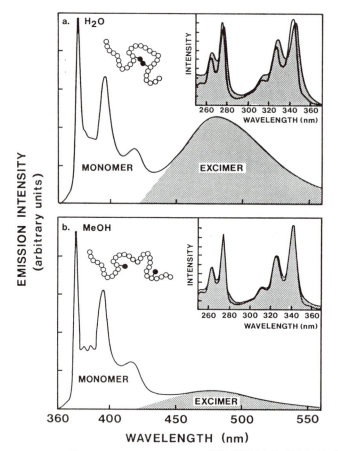

Figure 3:   Fluorescence spectra of PNIPAM-Py/200 (0.05 g L⁻¹) in water and in methanol; λ_{exc} = 330 nm.

being defined as the interchromophoric distance at which one half of the donor molecules decay by energy transfer and one half by the usual radiative and non-radiative processes.   With λ_{exc} = 290 nm, a wavelength at which most of the light is absorbed by naphthalene, one can detect both the direct emission from excited naphthalene and the emission from pyrene excited through NRET from Na*.   Thus under circumstances where the pyrene and naphthalene groups are in close enough proximity to satisfy the NRET requirements, excitation at 290 nm will result in a complex emission consisting of the emission from Na* (310 to 400 nm) and the emission from Py* excited by transfer of energy from Na*.   By comparing the emission of the mixed solution to the spectra of the solutions of each polymer (Figure 4), one observes a significant quenching of the naphthalene emission and a dramatic enhancement of the pyrene monomer emission, even though the concentration of each chromophore is the same for corresponding solutions.   The high efficiency of NRET between Na* and Py is yet another indication of the existence in solution of interpolymeric micellar aggregates bringing together in close proximity the two interacting chromophores.   Note that the emission of the doubly labeled polymer, PNIPAM-C_{18}Na/360-C_{18}Py/2000,   in   water   also   exhibits   features characteristic of efficient NRET.   In this case indeed interactions between Na* and Py linked to the same polymer cannot be distinguished from those among chromophores attached to two different chains.

It is important to include here the contrasting results obtained with mixed solutions of either PNIPAM-Na/27 and PNIPAM-Py/200 (9), or PNIPAM-C_{18}Na/200 and PNIPAM-Py/200: in neither case was it possible to detect any significant extent of NRET between Na* and Py. Thus it appears that NIPAM polymers that do not carry a few long alkyl chains along their backbone are not incorporated in the polymeric micelles formed by the amphiphilic copolymers. Moreover, even though the dyes attached to the polymers and in particular pyrene, are hydrophobic in nature, they are not able to cause PNIPAM to form interpolymeric micellar structures. Such effects, described in detail elsewhere, will be exploited here.

Properties of Aqueous Solutions of PNIPAM and its Hydrophobic Copolymers.   PNIPAM and its lightly labeled derivatives used here do not aggregate in cold water in the dilute regime probed by our experiments. Evidence in support of this situation stems not only from the spectroscopic data reviewed in the preceding section, but also from light scattering experiments performed in several laboratories (7).   In contrast, amphiphilic NIPAM copolymers form polymeric micellar structures with the following salient features: 1) they consist of a highly viscous hydrophobic core formed by the alkyl substituents and a loose corona made up by the polymer main chain; 2) they exist in extremely dilute solutions and they always involve several polymer chains; 3) they seem to be stable indefinitely in cold water, but they are disrupted severely when the solutions are heated above their LCST: the original hydrophobic core is destroyed and the hydrophobic groups are accommodated as isolated entities within the separated polymer-rich phase.   These conclusions were drawn from a combination of results from experiments with extrinsic fluorescent probes, pyrene   (20), perylene (9), and dipyme (21), with pyrene-labeled copolymers, and with mixtures of pyrene- and naphthalene-labeled copolymers (22).

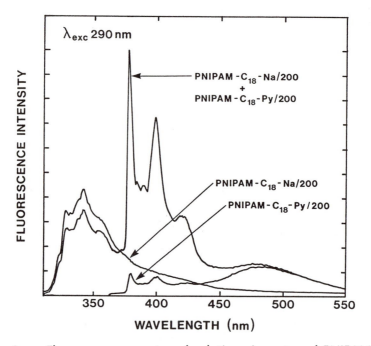

Figure 4:   Fluorescence spectra of solutions in water of PNIPAM-C$_{18}$Na/200, PNIPAM-C$_{18}$Py/200, and of a mixture of the two polymers; total polymer concentration: 0.39 g L^{-1}, λ_{exc} = 290 nm.

They corroborate conclusions reached by Schild and D. Tirrell from an independent study of similar amphiphilic NIPAM copolymers (23).

Interaction of the Polymers and Liposomes.

Experimental Design.   Dimyristoylphosphatidylcholine   (DMPC)   and distearoylphosphatidylcholine (DSPC) liposomes were prepared by dispersion of the phospholipids in water and subsequent sonication of the suspensions. Under these experimental conditions single unilamellar vesicles (SUV) form predominantly. Their diameters, measured by quasi-elastic light scattering (QELS) ranged between 90 and 120 nm in all the preparations. The two lecithins, DMPC and DSPC (Figure 5), were chosen to take advantage of two favorable circumstances: 1. their liposomes exhibit similar surface charge characteristics, as dictated by the polar choline head group; and 2. the temperature range between their respective transition temperatures, T_c, (DMPC: 23°C, DSPC: 54°C) (24) encompasses the LCST of PNIPAM (32°C). This latter fact was a key consideration in the design of the experiments. The temperature dependence of the structure and fluidity of lipidic bilayers is well-documented (25). Below a characteristic temperature, T_c, the phase transition temperature, liposomes exist in the solid-analogous phase in which the alkyl chains are crystallized in an all-trans configuration. There is practically no lateral diffusion within the membrane in this phase. As the temperature is raised through its T_c the bilayer passes from this tightly ordered "solid" or "gel" phase to a liquid-crystal phase, where freedom of movement of individual chains is higher. The alkyl groups are able to adopt conformations other than the straight all-trans configuration and to undergo rotational movements. This tends to create void volumes and to expand the area occupied by the chains and to allow lateral diffusion within the plane of the membrane. The phase transition temperature of DMPC liposomes is such that experiments carried out at room temperature (25°C) monitor the interactions of polymers and a fluid membrane. Experiments with DSPC at room temperature probe the interactions between the polymers and a rigid membrane.

About: Another important consideration in the design of the experiments had to do with the relative concentrations of polymers and liposomes in the samples under study. In all cases, concentrations of polymers and lipids were chosen such that there was a large excess of liposomes relative to polymers. An estimated 2 to 7 liposomes were left intact for each polymer-coated liposome, assuming that all the liposomes were unilamellar and that each SUV (average diameter: 100 nm) contains 2850 lipids (26). Under these conditions one can monitor easily changes in the properties of the polymers, but, since only a few liposomes are coated with polymer, measurements aimed at detecting intrinsic properties of the liposomes, such as calorimetric determinations of phase transitions will yield values averaged over treated and intact vesicles. Most experiments were carried out with the amphiphilic copolymers, but we examined also the adsorption of PNIPAM and its labeled derivatives. Comparisons will be drawn between the two sets of polymers in an attempt to highlight on the one hand, the influence of the degree association of polymers before their encounter with liposomes and, on

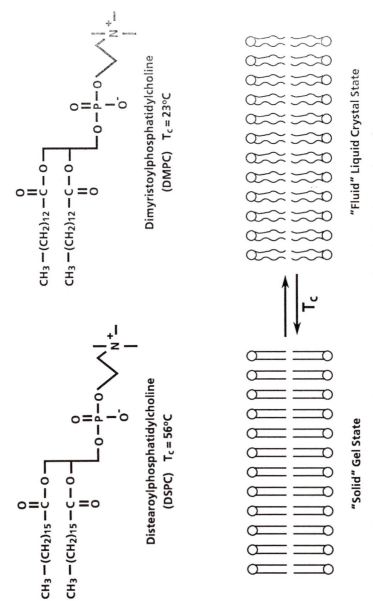

Distearoylphosphatidylcholine
(DSPC) $T_c = 56°C$

Dimyristoylphosphatidylcholine
(DMPC) $T_c = 23°C$

"Solid" Gel State

"Fluid" Liquid Crystal State

Figure 5:   Chemical structure of the phospholipids used in this study and schematic representation of the 'solid' phase and 'fluid' phase of lipidic bilayers.

the other hand, the role played by the octadecyl chains in anchoring individual polymer chains onto the liposome bilayer.

Microcalorimetry. Differential scanning calorimetry (DSC), a well established technique in the study of the phase transitions of liposomes (27), has proven extremely useful in detecting the interactions between liposomes and proteins (28) or synthetic polymers (29). Recently DSC has been applied also to evaluate the thermodynamic parameters associated with the phase separation which occurs at the LCST of polymer solutions, such as aqueous PNIPAM (30). Enticed by these reports, we performed a series of DSC scans on the solutions employed in the fluorescence measurements. In these samples the polymer concentrations were quite low, and in consequence the DSC signal was low and rather noisy. Also, as the liposomes were in large excess relative to the polymers, only changes in the transition associated with the LCST are meaningful here. The data corresponding to the lipid bilayer phase transitions represent some average over coated and uncoated liposomes. DSC curves of two polymer/DSPC systems are shown in Figure 6. The trace corresponding to the NIPAM homopolymer/DSPC system exhibits two endotherm transitions, one at 32.3°C attributed to the phase transition of the polymer, and a second one at 56°C, assigned to the gel-liquid transition of the lipid bilayer. The LCST transition of the homopolymer is not affected by the presence of the liposomes, as reported also in the case of mixtures of PNIPAM and liposomes of dipalmitoyl L-α-phophatidylcholine, phosphatidic acid, or phosphatidyl glycerol (31). The DSC curve of the liposomes in the presence of a mixture of the hydrophobically-modified copolymers, the [PNIPAM-C_{18}Na/200 and PNIPAM-C_{18}Py/200]/DSPC system, does not exhibit the signal corresponding to the LCST of the copolymer mixture (32.3°C). DSC curves of polymer/DMPC systems exhibited a different phenomenon: the endotherm at 32.3°C due to the LCST of the polymers was observed both in the case of PNIPAM/ DMPC and in the case of the amphiphilic copolymers/DMPC system, although the signal due to the hydrophobically modified polymers is weaker and broader in presence of liposomes than in solution.

These preliminary microcalorimetric measurements suggest that there is apparently no interaction between the homopolymer and the liposomes, since the LCST transition is unaffected by the presence of liposomes. However the signal corresponding to the LCST of the hydrophobically-modified polymers is altered in the presence of liposomes, in terms of both intensity and broadness. This observation is taken as a clue that these polymers interact with the liposomes. Particularly revealing is the fact that for liposomes in their fluid state (DMPC) an LCST transition can still be detected, but this transition seems to disappear for liposomes in their solid-analogous state (DSPC). This dependence of the polymer/liposome interactions on the membrane phase will be put on much firmer grounds by the fluorescence experiments described next. Indeed further microcalorimetric measurements with fully coated liposomes systems need to be performed in order to assess the consequences of adsorbing amphiphilic NIPAM copolymers on the bilayer phase transitions. However the absence in the DSC traces of the signal due to the polymer transition in the case of the amphiphilic copolymers but not the homopolymer

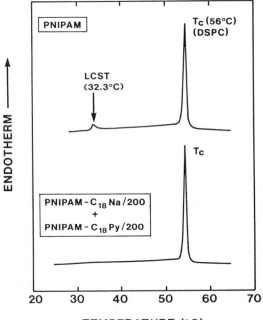

Figure 6:   Microcalorimetric endotherms for aqueous DSPC/polymer samples: Top: PNIPAM, bottom: PNIPAM-C_{18}Na/200 and PNIPAM-C_{18}Py/200.

highlights the key role of the octadecyl groups in guiding the polymer/liposomes interactions.

Spectral Response of the Labeled Polymers During the Adsorption Process

Evidence from the Excimer Emission of Pyrene-labeled Polymers. The addition of a solution of PNIPAM-C_{18}Py/200 to   DMPC or DSPC liposomes triggered remarkable changes in the emission of the labeled polymer (Figure 7) (4). Most noticeable was the sharp decrease in the pyrene excimer emission, indicative of   an increase in the interchromophoric separation beyond the distance probed by the excimer.   When increasing amounts of the   labeled copolymer were added to a liposome suspension of constant concentration, a polymer concentration was reached at which the ratio I_E/I_M increased again. This signals the point at which the surface of all liposomes are saturated with adsorbed polymer chains. Interestingly the same response was triggered by the addition of DSPC liposomes to a solution of PNIPAM-Py/200, the polymer labeled with pyrene, but devoid of octadecyl groups.   Hence this polymer too is adsorbed until saturation on the surface of liposomes (Table IIa).

Evidence from Non-Radiative Energy Transfer Experiments. The pyrene excimer experiments do not allow one to distinguish the interactions of pyrenes attached to the same polymer chain from those among pyrenes belonging to two different polymers.   Interpolymeric interactions can be detected by monitoring NRET between chromophores attached to two distinct polymers, PNIPAM-C_{18}-Na/200 and PNIPAM-C_{18}-Py/200 for example.   Thus it is possible to follow changes in distances between polymers as they interact with the liposomes. The extent of Na* to Py NRET is reported here in terms of the ratio, I_{Na}/I_{Py}, of the intensity of the naphthalene emission (343 nm) to that of the pyrene emission (378 nm). This value provides a convenient scale for the comparison of the various systems, but it cannot be related easily to the real efficiency of the energy transfer.   A low I_{Na}/I_{Py} value corresponds to situations where the extent of NRET is high (Table IIb).

Addition of a suspension of DMPC or DSPC liposomes to a mixed solution of hydrophobically-modified polymers labeled with either naphthalene or pyrene, PNIPAM-C_{18}Na/200 and PNIPAM-C_{18}Py/200, caused a decrease of the pyrene emission intensity and a concurrent increase of the naphthalene emission intensity (Figure 8).   This observation is diagnostic of a decrease in the extent of NRET between Na* and Py, implying a lengthening, on average, of the interchromophoric distances.   Changes occurred immediately upon mixing of the polymer solution and the liposome suspension, but it took about two hours for the emission to remain constant with time.   The decrease in the extent of NRET was more pronounced, it seems, when the polymers entered in contact with DSPC liposomes than with the DMPC liposomes (Table III), stressing the importance of the fluidity of the membrane on its interaction with the polymers.   Addition of liposomes to a solution of the doubly labeled amphiphilic polymer PNIPAM-C_{18}Na/360-C_{18}Py/2000   (10 mg L^{-1}) also caused a marked decrease in the extent of NRET, as evidenced by the significant increase in the ratio I_{Na}/I_{Py}. As observed with the mixed polymers solutions the decrease in NRET was more pronounced upon addition of DSPC

Table IIa. Compositions of the samples examined: Pyrene excimer emission

Liposome Component:		Polymer Component:	
Liposome (1 g L⁻¹)	Size (nm)	Polymer structure	Concentration (g L⁻¹)
DMPC	110-130	PNIPAM-C$_{18}$-Py/200	66 x 10⁻³
DSPC	110-120	PNIPAM-C$_{18}$-Py/200	66 x 10⁻³
DSPC	110-120	PNIPAM-Py/200	66 x 10⁻³

Table IIb. Compositions of the samples examined: Non-radiative energy transfer experiments

Liposome (1 g L⁻¹)	Size (nm)	Polymer	Concentration g L⁻¹ (Na:Py ratio)	
DMPC	110-130	PNIPAM-C$_{18}$Na/200 + PNIPAM-C$_{18}$Py/200	(5:1)	
DSPC	110 -120	PNIPAM-C$_{18}$Na/200 + PNIPAM-C$_{18}$Py/200	(5:1)	
DMPC	110-130	PNIPAM-C$_{18}$Na/360-C$_{18}$Py/2000		
DSPC	110-120	PNIPAM-C$_{18}$Na/360-C$_{18}$Py/2000		
DSPC	110-120	PNIPAM-Na/27 + PNIPAM-Py/200	24 x 10⁻³	(5:1)
DSPC	110-120	PNIPAM-C$_{18}$Na/200 + PNIPAM-Py/200	39 x 10⁻³	(5:1)
DSPC	110-120	1. PNIPAM-C$_{18}$Na/200 (preadsorbed) 2. PNIPAM-Py/200	39 x 10⁻³	(5:1)

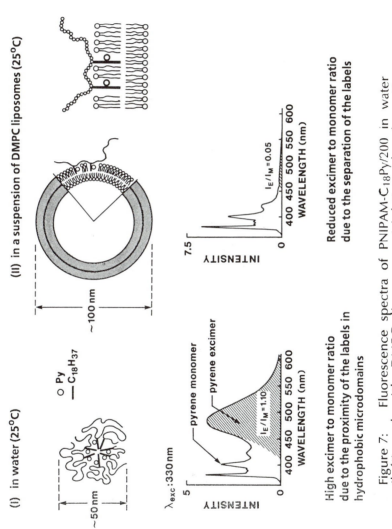

Figure 7: Fluorescence spectra of PNIPAM-C_{18}Py/200 in water (left) and with DMPC liposomes (right) and schematic representations of the interactions; 25°C, $\lambda_{exc} = 330$ nm.

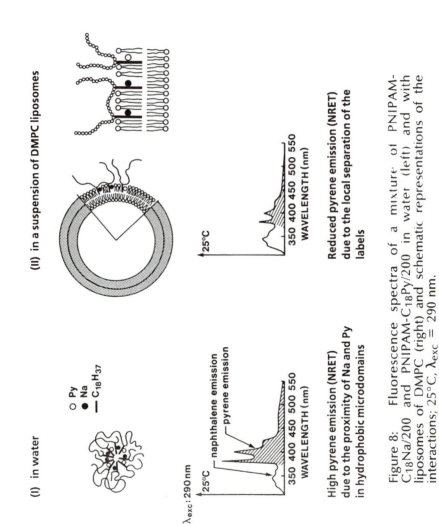

Figure 8: Fluorescence spectra of a mixture of PNIPAM-C_{18}Na/200 and PNIPAM-C_{18}Py/200 in water (left) and with liposomes of DMPC (right) and schematic representations of the interactions; 25°C, λ_{exc} = 290 nm.

Table III. Photophysical Parameters of the Polymers in Solution and in the Presence of Liposomes

Polymer	Water		DSPC Liposomes		DMPC Liposomes	
	I_E/I_M [a]	I_{Na}/I_{Py} [b]	I_E/I_M [a]	I_{Na}/I_{Py} [b]	I_E/I_M [a]	I_{Na}/I_{Py} [b]
PNIPAM-C$_{18}$Py/200	1.10	--	0.04	--	0.04	--
PNIPAM-Py/200	0.40	--	0.02	--	0.02	--
PNIPAM-C$_{18}$Na/360-C$_{18}$Py/2000	0.06	0.02	0.02	0.40	0.02	0.32
P-C$_{18}$Na/200 + P-C$_{18}$Py/200	0.80	0.20	0.02	0.90	0.02	0.67
P-Na/27 + P-Py/200	0.40	0.90	0.02	0.41	0.02	c
P-C$_{18}$Na/200 + P-Py/200	0.40	1.7	0.02	1.20	0.02	c

a: Excimer to monomer intensities ratio for direct pyrene emission (λ_{exc} 330 nm), estimated error: ± 0.02;
b: Naphthalene to pyrene emission intensities ratio (λ_{exc} 290 nm), estimated error: ± 0.02;
c. not measured

liposomes. Addition of the terpolymer to a DMPC liposome suspension (1.0 g L-1, diameter: 100 nm) triggered no change in the I_{Na}/I_{Py} up to a polymer concentration of 60 mg L-1. At this point further addition of polymer resulted in a large decrease of the ratio, signalling the presence of free polymeric micelles together with polymer saturated vesicles.

To assess the role of the octadecyl groups during the adsorption process experiments were run with labeled NIPAM polymers devoided of octadecyl groups. Changes in NRET were recorded as liposomes were placed in contact with a mixed solution of a naphthalene labeled polymer, PNIPAM-Na/27, and a pyrene-labeled polymer, PNIPAM-Py/200

The fluorescence spectrum (λ_{exc} 290 nm) of a mixed solution of PNIPAM-Na/27 and PNIPAM-Py/200 (Figure 9a) consists of a contribution from the naphthalene emission (310 to 400 nm) and a weak emission (378 to 600 nm) attributed to pyrene monomer and excimer emissions. This emission (total intensity and ratio I_E/I_M) is identical to the spectrum of a solution of PNIPAM-Py/200 under the same conditions of chromophore concentration and excitation (Figure 9, top of letf section, hatched area). It is assigned therefore to the emission of directly excited pyrene and not to the NRET process. The absence of interactions between the naphthalene and pyrene chromophores confirms that the polymers do not aggregate in water. Upon contact with DSPC liposomes small changes occur in the fluorescence spectrum (Figure 9, right section): most noticeable are an increase of the pyrene monomer emission and a small decrease of the naphthalene emission, indicating that the NRET process takes place now, although only to a small extent. Hence in this new situation interactions between chromophores from different chains are possible. The chromophores are brought into proximity, it is believed, upon adsorption of polymer chains onto the liposome surface. Since the liposomes are in large excess, only few interpolymeric contacts are made, hence the extent of energy transfer remains low. This type of measurement would not detect the eventual presence of polymer still dissolved in water. It is possible however to monitor the emission from PNIPAM-Py/200 by direct excitation of pyrene (λ_{exc} 330 nm), and to determine the consequences of liposome addition on the extent of pyrene excimer emission: we observed, as was the case for solutions of PNIPAM-Py/200 in the absence of liposomes (vide supra), a complete disappearance of the excimer emission, supporting evidence of adsorption of the polymers on the liposomes.

The "Molecular Accordion." This study of polymer/liposome systems focussed primarily on the response of the polymers, varying parameters such as the phase of the lipid bilayer and the degree of association of the polymers in water. Since the samples always contained a large excess of liposomes relative to the polymer, PNIPAM and its lightly labeled derivatives which do not aggregate in water, are expected to distribute themselves mostly as isolated chains on the surface of many liposomes. In contrast, amphiphilic copolymers of NIPAM in water form interpolymeric aggregates leaving only few, if any, isolated chains in solution.   Thus the adsorption proceeds via a diffusion-controlled encounter of vesicles and polymeric aggregates. Several polymer chains adsorb on a single vesicle. Reorganization of the polymers on the vesicle surface is expected to take place, with some degree of penetration of the octadecyl groups within the lipidic bilayer. Migration of the polymers from vesicle to vesicle is unfavored, since it would require for the

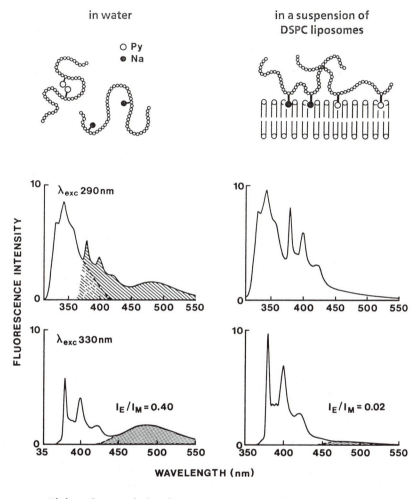

**High excimer emission from   Reduced excimer emission due to
intrapolymeric Py-Py dimers   the separation of the labels**

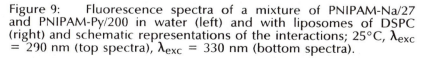

Figure 9:   Fluorescence spectra of a mixture of PNIPAM-Na/27 and PNIPAM-Py/200 in water (left) and with liposomes of DSPC (right) and schematic representations of the interactions; 25°C, λ_{exc} = 290 nm (top spectra), λ_{exc} = 330 nm (bottom spectra).

octadecyl substituents to exit the lipidic hydrophobic environment and relocate in bulk water. While insertion of octadecyl groups within a fluid membrane can be expected to be a facile process, details of the interactions of the amphiphilic copolymers and the DSPC liposomes below their T_c remain unclear at this point. It may be argued that the octadecyl groups, unable to insert themselves among the stiff lipid chains, remain located mostly in the outer layer of the liposomes.

In our first report on the 'molecular accordion' effect, we described temperature-dependent studies, using evidence from intensity changes in pyrene excimer emission from PNIPAM-C_{18}Py/200 (6). Raising the temperature of the DMPC/PNIPAM-C_{18}Py/200 system above 32°C, the LCST of the polymer, caused a small but significant enhancement of the excimer emission, an indication that the contacts between pyrene groups increased as the polymer chains collapsed. The process was thermoreversible. When the same experiment was repeated with DSPC liposomes which have a phase transition temperature well above the LCST of the polymer, this increase in the excimer emission could be detected only above 56°C, when the lipid bilayer was in the liquid-analogous phase. Moreover the process was not thermoreversible: cooling the system below 32°C did not bring about a recovery of the excimer. Although the polymer chains are expanded, the anchored chromophores are held separated from each other, as a result, we surmise, of the anchoring of the octadecyl chains within the rigid lipid bilayer. Temperature-dependent studies of polymer/liposome interactions using NRET between naphthalene and pyrene are under progress and will be described shortly.

Conclusions

In summary, the fluorescence experiments and the microcalorimetric measurements provide strong evidence for the following description of the interactions between amphiphilic NIPAM copolymers and liposomes: 1. the polymeric micellar structures formed in water are disrupted irreversibly when they enter in contact with liposome surfaces; 2. polymer chains reorganize on the surface of the vesicle in a slow process, which results in a separation of the octadecyl groups previously held in close contact within the polymeric micelles; 3. upon saturation coverage of the liposome surface, coated liposomes and free polymeric micellar structures coexist in solution; 4. the phase of the lipidic bilayer plays a role, it seems, during the adsorption process. This last observation, based primarily on preliminary DSC measurements, should be clarified by monitoring the spectroscopic response of the labeled polymer/liposome systems to changes in temperature. Also, a closer look at the time-dependence of the extent of NRET between the Na* and Py is essential in order to understand the reorganization process of the octadecyl as they encounter lipidic bilayers.

Acknowledgments. Financial support for this work was provided in part (H.R. and J. S.) by the Bundesministerium für Forschung und Technologie (BMFT) and by the Deutches Forschungsgemeninschaft (DFG). The DSC measurements were carried out in the laboratory of Professor E. Sackman (T. Universität, München). We thank Professor Sackman for the use of his equipment and for helpful discussions.

Literature Cited

1. For reviews, see for example: *Liposomes as Drug Carriers;* Schmidt, K. H., Ed; Georg Thieme, Verlag, Stuttgart, 1981; *Liposome Technology;* Gregoriadis, G, Ed.; CRC Press, Boca Raton, Florida, 1983; Vol. 2.

2. Yasuda, T.; Tadakuma, T;, Pierce, C. W.; Kinsky, S. C. *J. Immunol.* **1979**, *123*, 1535.

3. Caride, V. J.; Sostman, H. D.; Twickler, J.; Zacharis, H.; Orphanoudakis, S. C.; Jaffe, C. C. *Invest. Radiol.* **1982**, *17* ,3815.

4. Ringsdorf, H.; Schlarb, J.; Venzmer, J. *Angew. Chem. Int. Ed.* **1988**, *127*, 113.

5. Özden, M. Y.; Hasirci, V. N. *British Pol. J.* **1990**, *23*, 229; Baskin, A.; Rosilio, V.; Albrecht, G.; Sunamoto, J. *J. Colloid Interf. Sci.* **1991**, *145*, 502.; Sunamoto, J.; Sato, T. Hirota, M.; Fukushima, K.; Hiratano, K.; Hara, K. *Biochim. Biophys. Acta* **1987**, *898*, 323; Ishihara, K.; Nakabayashi, N. *J. Polymer Sci.:Polymer Chem.* **1991**, *29*, 831.

6. Ringsdorf, H.; Venzmer, J.; Winnik, F. M. *Angew. Chem. Int. Ed. Engl.* **1991**, *30*, 315.

7. Binkert, Th.; Oberreich, J.; Meewes, M.; Nyffenegger, R.; Ricka, J. *Macromolecules* **1991**, *24*, 5806; Fujishige, S.; Kubota, K.; Ando, I. *J. Phys. Chem.* **1989**, *93*, 3311; Yamamoto, I.; Iwasaki, K.; Hirotsu, S. *J. Phys. Soc. Jpn.* **1989**, *58*, 210.

8. Winnik, F. M. *Macromolecules* **1990**, *23*, 233.

9. Winnik, F. M. *Polymer* **1990**, *31*, 2125.

10. Ringsdorf, H.; Venzmer, J.; Winnik, F. M. *Macromolecules* **1991**, *24*, 1678.

11. Ringsdorf, H.; Simon, J.; Winnik, F. M. *Macromolecules* **1992**, *25*, 5353.

12. New, R. R. C. In *Liposomes, a Practical Approach*, R. R. C. New, Ed.; IRL Press, Oxford University Press, New York, NY, 1990; chapter 2.

13. Birks, J. B. *Photophysics of Aromatic Molecules;* Wiley-Interscience: London, 1979; chapter 7.

14. Winnik, F. M. ; Winnik, M. A.; Tazuke, S.; Ober, C. K. *Macromolecules* **1987**, *20*, 38; Oyama, H. T.; Tang, W.; Frank, C. W. *Macromolecules* **1987**, *20*, 477, 1839; Winnik, F. M.; Tamai, N.; Yonezawa, J.; Nishimura, Y.; Yamazaki, I. *J. Phys. Chem.* **1992**, *96*, 1967.

15. Förster, T. *Discuss. Faraday Soc.* **1959**, *27*, 7.

16. Lakowicz, J. R. *Principles of Fluorescence Spectroscopy*, Plenum Press, New York, NY, **1983**, Chapter 10.

17. Nagata, I.; Morawetz, H. *Macromolecules* **1981**, *14*, 87.

18. For a review, see for example: Kleinfeld, A. In *Spectroscopic Membrane Probes*, Loew, L. M., Ed.; CRC Press, Boca Raton, Florida, Vol. 1, 1988, chapter 4, *and references therein.*

19. See for example, Morris, S. J.; Bradley, D.; Gibson, C. C.; Smith, P. D.; Blumenthal, R. In *Spectroscopic Membrane Probes*, Loew, L. M., Ed.; CRC Press, Boca Raton, Florida, Vol. 1, 1988, chapter 7, and references therein.

20. Winnik, F. M.;   Ringsdorf, H.; and Venzmer, J. *Macromolecules* **1990**, *23*, 2615.

21. Winnik, F. M.; Winnik, M. A.; Ringsdorf, H.; Venzmer, J. *J. Phys. Chem.* **1991**, *95*, 2583.

22. Ringsdorf, H.; Simon, J.; Winnik, F. M. *Macromolecules* **1992**, *25*, 5353.

23. Schild, H. G.; Tirrell, D. A. *Langmuir* **1991**, *7*, 1319.

24. Mabrey, S.; Sturtevant, J. M. *Proc. Natl. Acad. Sci. USA* **1976,** *73*, 3862.

25. Chapman, D. *Quart. Rev. Biophys.* **1975**, *8*, 185.

26. Watts, A.; Marsh, D.; Knowles, P. F. *Biochemistry* **1978**, *17*, 1792.

27. Ladbrooke, B. D.; Chapman, D. *Chem. Phys. Lipids* **1969,** *3*, 304.

28. Papahadjopoulos, D. *J. Colloid Interf. Sci.*, **1977**, *58*, 459.

29. Schroeder, U. K. O.; Tirrell, D. A. *Macromolecules* **1989**, *22, 765.

30. Schild, H. G.; Tirrell, D. A. *J. Phys. Chem.* **1990**, *94*, 4356.

31. Schild, H. G. *Progr. Poly. Sci.* **1992,** *17*, 163.

Chapter 18

Interaction of Water-Soluble Polymers with Dilute Lamellar Surfactants

J. Frederick Hessel[1,2], Richard P. Gursky[2], and Mohamed S. El-Aasser[1]

[1]Department of Chemical Engineering and Emulsion Polymers Institute,
Lehigh University, Bethlehem, PA 18015
[2]Unilever Research, U.S., Edgewater, NJ 07020

The effect of water soluble polymers [poly (ethylene glycol) or poly (sodium acrylate)] on the morphology and properties of vesicles prepared from sodium dodecyl sulfate (SDS) and dodecanol ($C_{12}OH$) has been studied with rheological measurements, microscopy, and differential scanning calorimetry. The SDS/$C_{12}OH$ system forms lamellar gel (L_β) phase vesicles in dilute aqueous solutions (< 15 mM) and the addition of water soluble polymers has a significant effect on the properties of the system. Exclusion of the polymers from the intra-lamellar water layers results in an osmotic compression of the vesicles and a decrease in the phase volume and viscosity of the dispersion. The exclusion of polymers was found to depend on the polymer molecular weight and the temperature of the polymer addition. Polymers added below the gel-to-liquid crystal phase transition temperature were excluded from the vesicle independent of molecular weight. Polymer exclusion when the polymers were added above the gel-to-liquid crystalline transition depended on polymer molecular weight. When the polymer molecular weight was increased such that the radius of gyration of the polymer coil is half the water layer thickness, then the polymer was excluded and osmotic compression was observed. Electrostatic interactions dominated the behavior of the poly (sodium acrylate) systems. Depletion flocculation and fusion of the vesicles was observed at increased polymer concentrations, but no direct interaction between the polymers and the bilayers was observed.

The interaction of water soluble polymers with lamellar dispersions has implications in many practical applications such as consumer products, drug delivery, bioseparations, and industrial processes. The osmotic stress method for studying membrane biophysics involves the addition of water soluble polymers such as poly (ethylene glycols) (PEGs) to phospholipid vesicles and measuring the changes in

0097–6156/93/0532–0241$06.00/0

bilayer dimensions (*1*). The exclusion of PEGs from the intra-lamellar water layers in multi-lamellar vesicles has been reported for phospholipid dispersions (*2*) and mixed surfactant systems used in detergent compositions (van de Pas, J.C., Unilever Research, Vlaardingen, Holland, personal communication, 1991).

This exclusion of polymers from the interior of the vesicles results in an osmotic compression of the water layers and a decrease in the water layer thickness and lamellar phase volume. This effect allows the control of bulk properties such as viscosity and also provides a probe of water layer dimensions in lamellar dispersions. The lamellar surfactant system used in this study is the sodium dodecyl sulfate (SDS)/dodecanol ($C_{12}OH$)/water system that has been used to prepare submicron diameter emulsions (miniemulsions) from monomers for emulsion polymerization (*3*) and for the preparation of artificial latexes by direct emulsification of polymer solutions such as ethyl cellulose (*4*). This surfactant system forms lamellar dispersions (vesicles) in water at very low surfactant concentrations (< 13 mM).

The thermotropic and lyotropic phase behavior of these mixed SDS/alkanol systems has been found to be governed by the geometric packing constraints imposed by the structure of the surfactant and cosurfactant. At an average headgroup area (A_h) < 0.47 nm^2, the lamellar phase was observed, as predicted by the geometric packing constraint model for a C_{12} hydrocarbon chain (*5*). A maximum in the gel-to-liquid crystalline transition temperature (T_m) was observed with compositions where A_h matches the area occupied by the alkyl chains ($A_h = 0.25$ nm^2). The optimum packing results in a maximum in the van der Waals attractive forces and T_m.

These dilute compositions have been imaged by optical and electron microscopy and found to consist primarily of uni- and multi-lamellar vesicles with a diameter of 1.0 μm. One objective of this study was to confirm the closed bilayer morphology observed by microscopic techniques. If the system was composed of continuous bilayers rather than closed vesicles, then no osmotic compression would be observed. A second objective was to estimate the water layer thickness of the vesicles by varying the polymer molecular weight and dimensions and compare the estimated water layer thickness measured by small angle neutron scattering (SANS) experiments. A final objective was to study and characterize the nature of the interaction of water-soluble polymers with dilute lamellar surfactant systems by varying the polymer structure (anionic vs. nonionic and molecular weight).

Materials and Methods

The SDS and dodecanol ($C_{12}OH$) used in this study were obtained from Aldrich Chemical, both are reported to be 98% pure. Poly (ethylene glycols) (PEGs) of various molecular weights were supplied by Union Carbide (Carbowax 600 to 20,000). Poly (vinyl alcohol) (PVA; 98% hydrolyzed; MW = 14 to 20K) was supplied by Air Products under the tradename Airvol 103. Low molecular weight poly (sodium acrylate) (PAA; MW = 5.5K) was supplied by Alco Chemical under the tradename Alco 602N and high molecular weight PAA (50K) was supplied by BASF under the tradename Sokalan PA-50. All surfactants and polymers were used as received.

Samples were prepared by mixing the SDS, $C_{12}OH$, and distilled-deionized water at 60 °C for two hours and cooling to ambient temperature with stirring. The lamellar phase composition was fixed at SDS/$C_{12}OH$ = 1/2 weight ratio, 3 weight %. The polymers were added under two different conditions, above (T > T_m) and below (T < T_m) the gel-to-liquid crystalline transition temperature (T_m). The T < T_m samples were prepared by mixing the prepared lamellar dispersion and aqueous polymer solution at ambient temperatures for two hours. The T > T_m samples were prepared by mixing the lamellar dispersion and polymer for two hours at 60 °C and cooling to ambient temperatures with stirring.

Rheological measurements were performed on a Haake RV20 with an MVII concentric cylinder geometry. The temperature was controlled at 25 °C and the shear stress measured as a function of shear rate (1 to 400 s^{-1}). The shear rate was linearly varied from 0 to 400 s^{-1} over 10 minutes.

Small angle neutron scattering (SANS) experiments were conducted on the NIST 30m instrument. The wavelength was 0.6 nm and the sample thickness was 1 mm. The scattered intensity was measured at scattering wave vectors (q) between 0.01 and 0.75 nm^{-1}. Raw data was corrected for background and empty cell scattering and converted to absolute intensity by scaling relative to a polystyrene standard. The repeat distance was measured by the position of the maximum in the scattered intensity and the bilayer thickness calculated using the Gunier approximation (at low q), which calculates a characteristic dimension (6).

Sample morphology was characterized by optical and electron microscopy. Video enhanced microscopy (VEM) was performed with a Zeiss Axiphot microscope with a 100X oil immersion objective. Images were enhanced with a Hammamatsu C2400/Argus 10 Image Processor and the images were recorded with a Sony UP-5000P video printer. Freeze fracture transmission electron microscopy (FFTEM) samples were prepared by freezing a supported drop of solution in a Freon 22 bath at -160 to -170 °C, and then fracturing the sample under vacuum. The samples were shadowed with platinum at an angle of 45° to a thickness < 10 nm and carbon masked at 90° to a thickness of 20 nm. Replicates were imaged with a Jeol 100C TEM operated at 80 kV.

Thermal phase transition temperatures (T_m) and enthalpies (ΔH) were measured with a DuPont DSC 2910 at a heating rate of 2 °C/min. A 10 mg sample was loaded in a sealed aluminum pan and pre-equilibrated at 5 °C for 10 minutes prior to heating. The enthalpy of the transition was calculated from the area under the curve using an indium standard.

Results and Discussion

Vesicle Morphology. An important characteristic of the lamellar surfactant systems is the water layer thickness. The SANS experiments found that the hydrocarbon layer thickness was 3.3 nm and the water layers had a distribution of thicknesses between 30 and 40 nm for the 1/2 (w/w) SDS/$C_{12}OH$ (3 wt.%) sample.

Based on these dimensions, the high lamellar phase volume and the microscopic observations, a schematic model of the vesicles was constructed and is shown in Figure 1. The 1/2 (w/w) SDS/C_{12}OH (3 wt. %) composition consists predominately of multi-lamellar vesicles of 1 μm diameter composed of 3-4 alternating bilayers and water layers surrounding a large central water cavity. A FFTEM image of a sample of this composition is shown in Figures 2A and 3A.

The addition of water soluble polymers had a significant effect on the vesicle morphology depending on polymer type, molecular weight, and temperature of addition. The results are summarized in Table I.   When the polymers were added

Figure 1. Schematic model of lamellar phase vesicle.

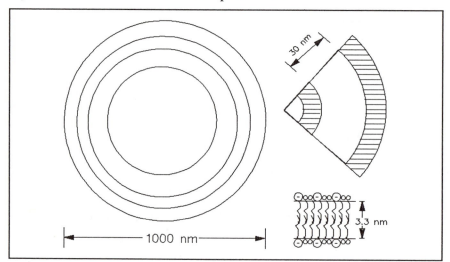

Table I. Effect of Water Soluble Polymers on Vesicle Morphology

	T>Tm	T<Tm
low MW	spherical	fibrous
	PAA: flocculated phase separated 0.5%	PAA: flocculated phase separated 2.5%
	PEG: non–flocculated compressed (at c > 1 wt.%)	PEG: flocculated and fused 2.5%
high MW	spherical	
	PAA: flocculated and fused phase separated .5%	PAA: fibrous flocculated and fused phase separated 2.5%
	PEG: non–flocculated compressed	PEG: non–flocculated compressed

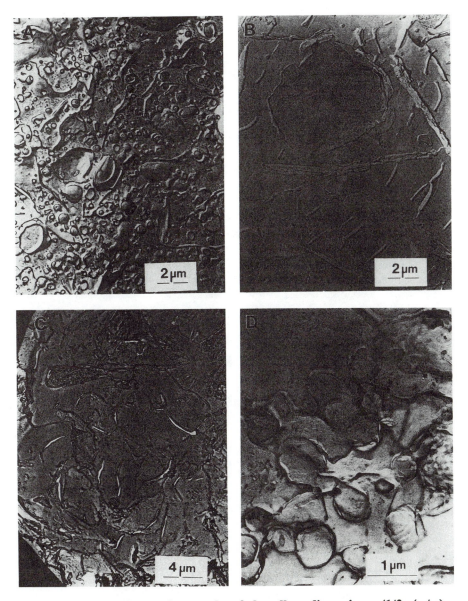

Figure 2. FFTEM micrograph of lamellar dispersions (1/2 (w/w) SDS/C$_{12}$OH, 3.0 wt.%) with added polymers: A) no polymer; B) +1.0% (1.8 mM PAA; MW = 5.5K; T > T$_m$; C) + 1.0% (1.8 mM) PAA; MW = 5.5K; T < T$_m$; and D) + 1.0% (0.2 mM) PAA; MW = 50K; T > T$_m$.

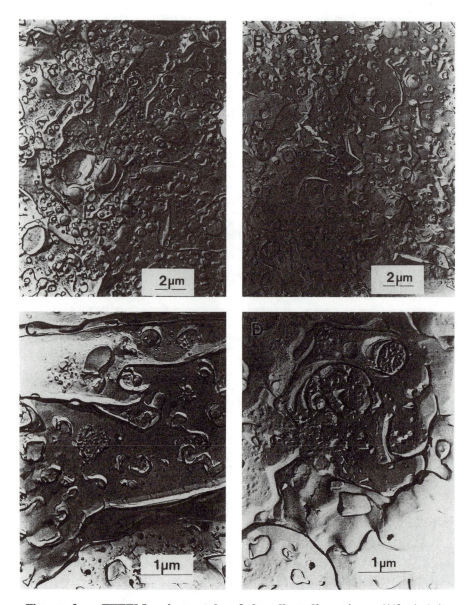

Figure 3.   FFTEM micrograph of lamellar dispersions (1/2 (w/w) SDS/C$_{12}$OH, 3.0 wt.%) with added polymers: A) no polymer; B) + 0.5 wt.% (1.5 mM) PEG 3350; T > T$_m$; C) + 2.5 wt.% (7.5 mM) PEG 3350; T < T$_m$; and D) + 2.5 wt.% (7.5 mM) PEG 3350; T > T$_m$.

at temperatures below (T < T_m) the gel to liquid crystalline transition, both the low molecular weight PEG and PAA resulted in a fibrous texture due to flocculation and fusion of the vesicles (see Figures 3C and 2C respectively). Samples containing PAA also phase separated into a surfactant rich phase and an aqueous phase due to screening of the repulsive electrostatic forces between the vesicles. The flocculation and fusion was primarily due to depletion flocculation, since it occurred with both PEG and PAA. Screening of electrostatic charges also contributed, since PAA flocculated the vesicles at lower concentrations than did PEG (see Figure 2C), but PAA was much more effective than simple electrolytes (i.e., NaCl) due to depletion flocculation.

High molecular weight PAA added at T < T_m also resulted in fusion of the vesicles and a fibrous texture, but the addition of PEG with MW = 20K did not. The difference between the high molecular weight PEG and PAA is due to electrostatics and possibly the difference in molecular weight (PEG = 20K and PAA = 50K). Low molecular weight PEG induced fusion but high molecular weight PEG did not, probably because on an equal weight basis there is a higher molar concentration of PEG 3350 (see Figures 3C and 3D).

When the polymers were added at T > T_m, there was no fibrous texture observed for either low or high molecular weight PEG or PAA. The PAA containing samples phase separated but at lower concentrations than the T < T_m samples. PEG 20K reduced the vesicle diameter from 1 μm to < 0.2 μm while PEG 3350 had little effect on vesicle diameter (see Figure 3B). These effects are caused by the steric exclusion of the polymer from the water layers which results in a concentraton gradient and osmotic compression of the vesicle. When the polymers were added at T > T_m, the fluid vesicles could break apart under shear and reform, incorporating the low molecular weight polymers. The high molecular weight polymers were not incorporated due to size exclusion.

Rheology of Lamellar Dispersions. The 1/2 SDS/C_{12}OH (3 wt.%) composition has a viscous, shear thinning rheology due to the high volume fraction of "hard spheres". A representative plot of viscosity versus shear rate from 1 to 400 s^{-1} for a sample containing 0.05 wt.% (0.15 mM) PEG 3350 added at T > T_m is shown in Figure 4. The reduction in vesicle diameter and phase volume due to osmotic compression by the polymers results in a significant reduction in viscosity.

The effect of polymer molecular weight on the viscosity when added at T < T_m is shown in Figure 5. The viscosities were measured at a shear rate of 100 s^{-1} and T = 25 °C. The viscosity ratio is the viscosity of the dispersion plus polymer to the viscosity of the dispersion. The viscosity of the lamellar dispersion is reduced by the addition of the nonionic polymers of all molecular weights and the ranking of effectiveness (i.e., 20K > PVA = 3350 > 600) is consistent with the expected osmotic pressure due to the second virial coefficient of the polymers. At an equal molar concentration the second virial coefficient and the osmotic pressure due to the concentration gradient increases with molecular weight.

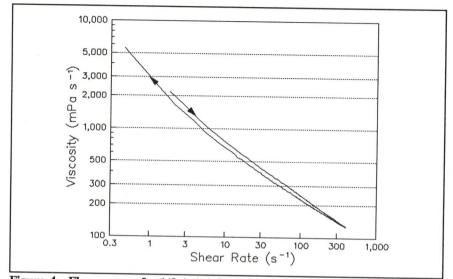

Figure 4. Flow curve of a 1/2 (w/w) SDS/C$_{12}$OH (3 wt.%) + 0.05 wt.% PEG 3350 T > T$_m$ sample measured at 25 °C.

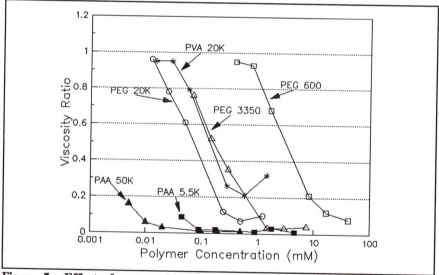

Figure 5. Effect of water soluble polymers on the viscosity ratio at 100 s^{-1} when the polymers were added at T < T$_m$.

The reduction in viscosity observed with PAA is much greater due to the screening of the intra- and inter-lamellar electrostatic repulsion. The water layer thickness of the vesicles is controlled by the electrostatic repulsion between SDS molecules in opposite bilayers and the addition of electrolyte screens this repulsion. This effect contributes to the reduction in vesicle diameter caused by the osmotic compression.

The calculated dimensions of the polymers are reported in Table II. The radii of gyration (R_gs) were calculated from viscosity/molecular weight relationships in a good solvent (7).

Table II. Dimensions of Polymers in Water

Polymer	R_g (nm)
PEG 600	1.9
PEG 3350	3.1
PEG 20K	7.7
PVA 17K	12.4
PAA 5.5K	5.4
PAA 50K	16.3

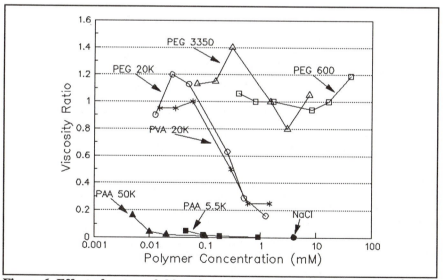

Figure 6. Effect of water-soluble polymers on the viscosity ratio at 100 s^{-1} when the polymers were added at T > T$_m$.

When the polymers were added at $T > T_m$ very different behavior was observed (see Figure 6). No reduction in viscosity was observed with the low molecular weight polymers (PEG 600 and 3350) and the reduction in viscosity with PEG 20K was observed at higher concentrations. At $T > T_m$ the vesicles can be broken under shear and reform resulting in the incorporation of polymer into the vesicle. PEG 600 and 3350 are not excluded from the water layers and no osmotic compression results. PEG 20K is excluded because R_g is half the water layer thickness, but since the polymer is polydisperse and there is a distribution of water layer thicknesses, some low molecular weight fractions are not excluded. This results in less osmotic compression than in the $T < T_m$ case. PVA (R_g = 12.4 nm) is entirely excluded under these conditions. Similar to the $T < T_m$ situation, electrostatics dominate the effects of PAA. The effect of simple electrolyte (NaCl) on the reduction in dispersion viscosity is shown for comparison (see Figure 6).

Effect of Polymers on Thermotropic Phase Behavior. In order to determine if the polymers had any direct effect on the bilayers, the gel-to-liquid crystalline phase transition temperature was measured for the samples. The gel-to-liquid crystalline phase transition temperature has been found to be a very sensitive measure of bilayer packing and order. If the polymers were directly interacting with the bilayer, it should be observed in the transition temperature and enthalpy.

Table III. Effect of Polymers on the Gel to Liquid Crystalline Transition

Polymer	Concentration (wt. %)	Temperature (°C)	T_m (°C)	ΔH (J/g)
None	0	60	38.8	1.0
PEG 3350	2.5	60	39.6	0.95
PEG 3350	2.5	25	39.7	0.97
PAA 5.5K	2.5	60	40.1	1.14
PAA 50K	1.0	25	39.4	1.00

The addition of polymers had only a slight effect on the transition temperature and enthalpy (see Table III), in spite of the significant morphological changes observed. A slight increase in the transition temperature was found with PEG 3350 added above or below the transition temperature (0.8 to 0.9 °C) which is an indication of increased bilayer packing. A larger increase was observed with 2.5% PAA with MW = 5.5K (1.3 °C), which is similar to that observed with simple electrolyte. This effect is probably due to a reduction in the repulsive electrostatic forces between SDS headgroups in the bilayer and an increase in packing and van der Waals attractive forces.

Effect of Polymers on Lamellar Dispersions. The effects of water-soluble polymers on lamellar dispersions are schematically summarized in Figure 7. Polymers are excluded from the vesicle water layers depending on the preparation temperature and polymer molecular weight. This exclusion results in a polymer concentration gradient and leads to an osmotic compression of the water layers which reduces the vesicle diameter and phase volume. With increased polymer concentration, depletion flocculation occurs due to exclusion of polymer from the vesicle surface. The flocculated vesicles undergo fusion resulting in a morphological change. Polyelectrolytes also screen the electrostatic repulsion between bilayers in the vesicle and between vesicles, and this effect dominates the osmotic compression. Screening of the repulsion between vesicles causes flocculation and phase separation of the lamellar phase. Screening of the charges between bilayers reduces the water layer thickness directly and enhances the osmotic compression caused by the polymer.

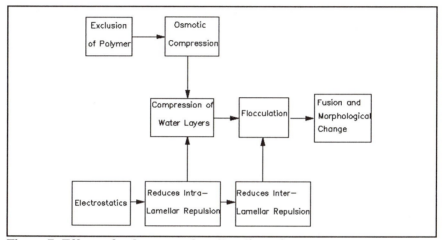

Figure 7. Effects of polymers on lamellar dispersions.

Summary and Conclusions

The effect of water-soluble polymers on the properties of dilute lamellar dispersions was studied and the primary effect was found to be due to the exclusion of polymer from the water layers. This exclusion results in an osmotic compression of the water layers and a reduction in the vesicle diameter, phase volume, and dispersion viscosity. Depletion flocculation leads to fusion and morphological changes in the dispersion. Polyelectrolytes screen the electrostatic repulsion between bilayers and vesicles and this effect is superimposed on the osmotic compression. There was no apparent direct interaction between the polymers and the bilayers observed.

Acknowledgments

Financial support for this program from Unilever Research, U.S. is greatly appreciated.

Literature Cited

1. Parseghian, V.A.; Fuller, N.L.; Rand, R.P. *Proc. Natl. Acad. Sci., USA* **1979**, 76, 2750.

2. Arnold, K.; Hermann, A.; Garwisch, K.; Pratsch, L.; *Stud. Biophys.* **1985**, 110, 135.

3. Ugelstad, J.; El-Aasser, M.S.; Vanderhoff, J.W.; *J. Polym. Sci., Poly. Lett.* **1973**, 111, 135.

4. Vanderhoff, J.W.; El-Aasser, M.S.; Ugelstad, J.; *US Patent 4,177,177* **1979.**

5. Mitchell, D.J.; Ninham, B.W.; *J. Chem. Soc., Faraday Trans. II* **1981**, 77, 601.

6. *Small Angle X-Ray Scattering*; Glatter, O.; Kratky, O., Eds.; Academic Press: New York, N.Y., 1975.

7. *The Polymer Handbook*; Bandrup, J.; Immergut, E.H., Eds; John Wiley: New York, N.Y., 1975.

RECEIVED May 10, 1993

Chapter 19

Characterization of Xanthan–Protein Complexes Obtained with Various Fermentation Conditions

Christine Noik, Frederic Monot, Daniel Ballerini,
and Jacqueline Lecourtier

Institut Français du Petrole, 1&4 avenue de Bois-Préau, B.P. 311, 92506
Rueil Malmaison Cédex, France

Fluid formulations used in the petroleum industry often contain synthetic polymers or biopolymers as thickening products. Biopolymers such as xanthan have different viscosifying properties, depending mainly on their chemical structure, acetate and pyruvate content and also on the purity of the samples, which can contain proteins.

The main aim of this study was to investigate the effects of the origin and amount of proteins on the aggregation of xanthan chains in solution by means of rheological measurements. Modifications of fermentation conditions led to variations of intrinsic viscosity, Huggins constant and pyruvate content of the xanthan produced.

The direct addition of protein-rich ingredients such as corn steep liquor or yeast extract to xanthan solutions also affected the rheological behavior of both diluted purified solutions and concentrated broth solutions. Aggregation effects on macromolecules chains and thixotropy phenomena have been observed. These effects are greatly dependent on salt content, pH, and nature of proteins, whether these proteins were added to xanthan solution or came from the fermentation medium.

Polymers with high molecular weight are often used for their viscosifying properties. Whatever the industrial application may be, the rheology of the macromolecular solution must be known and controlled. Xanthan is a polysaccharide often introduced

0097–6156/93/0532–0253$06.00/0

in fluid formulations for the petroleum industry, especially in chemical enhanced oil recovery processes or drilling operations (1,2).

Xanthan is an exo-cellular biopolymer produced by bacteria of the *Xanthomonas* genus. It is composed of glucose, mannose, glucuronic acid, pyruvate and acetate units. Its sugar composition is generally constant, but acetate and pyruvate contents may depend on the strain used (3,4) and on fermentation conditions such as nature and concentration of the carbon source, dissolved oxygen content and other nutrient limitations (5,6).

The characteristics of xanthan solutions depend on the polymer structure and also on the degree of purity of the samples produced by fermentation. It is a fact that it is very difficult to obtain xanthan in a pure form since some impurities, especially proteins, seem to be closely linked to it and some operations, such as for instance precipitation by an organic solvent, do not eliminate these contaminants. So, in order to control the viscosity, it is necessary to investigate the behavior of xanthan molecules in solution to have an idea of the aggregation state of polymer chains (7). The formation of aggregates has been found to be dependent on the presence and concentration of compounds such as proteins (8,9) and salts (10) which are issued from the fermentation medium. The mechanism of formation of xanthan-protein complexes and their nature are not known although electrostatic interactions are probable. Nevertheless other kinds of interactions can coexist. The presence of salts and physical treatments can also be involved in the formation of aggregates since microgels are essentially formed during downstream procedures (11).

After a preliminary analysis of the influence of the production mode on xanthan characteristics, this study was focussed on the effects of the origin and concentration of proteins on the state of aggregation of xanthan in solution estimated by rheological measurements. In a first while, purification treatments were applied to dilute polymer solutions in order to try to separate proteins from xanthan and thus to evaluate the amount and effect of proteins linked to the polysaccharide backbone. In the second phase of this study, the direct addition of proteins to dilute or concentrated xanthan solutions was analyzed, and the effects of salt, pH and origin of proteins were investigated.

MATERIALS and METHODS

Xanthan samples. Both commercial xanthans and samples produced in our laboratories (IFP-xanthans) were used in this study. Rhodopol 23 (Rhône-Poulenc, France) was supplied in a powder form and Flocon 4800 CX (Pfizer, USA) as a concentrated broth. The IFP-xanthan samples were referenced A, B or C according to their production mode.

Crude broths and powders were used in diluted or concentrated solutions. When diluted, the samples were centrifuged, ultrafiltered and sometimes treated as described.

The main characteristics of these products, determined by the methods mentioned below, are given in Table I. For all the commercial samples the molecular weight was found to be around 5.0 to 6.0 x10^6 daltons in 0.1 M NaCl. For samples

A, B and C, the molecular weight depended on the fermentation conditions as reported later in the text.

Production of xanthans A, B and C. *Xanthomonas campestris* DSM 1706 was grown at 25°C and pH 6.9 in a 6-liter LSL-Biolafitte fermentor. The aeration rate was about 0.05 vol/vol.min and the stirring rate was 600 rev/min. Preculture media were identical to fermentation media. Xanthan solution A was produced by batch fermentation on a complex medium consisting of 3 g.l^{-1} yeast extract, 3 g.l^{-1} malt extract, 5 g.l^{-1} peptone, 30 g.l^{-1} glucose, 5 g.l^{-1} K$_2$HPO$_4$ and 0.5 g.l^{-1} MgSO$_4$,7H$_2$O. Xanthan solutions B and C were obtained, respectively, by batch fermentation or sulfur-limited continous culture at a dilution rate of 0.02 h^{-1} on synthetic media. For xanthan solution B, the fermentation medium was composed of 0.1 g.l^{-1} yeast extract, 10 g.l^{-1} glucose, 0.5 g.l^{-1} (NH$_4$)$_2$SO$_4$, 2 g.l^{-1} K$_2$HPO$_4$, 0.5 ml.l^{-1} H$_2$SO$_4$ 85%, 0.2 g.l^{-1} Na$_2$SO$_4$, 0.1 g.l^{-1} MgSO$_4$,7H$_2$O, 0.023 g.l^{-1} CaCl$_2$, 0.007 g.l^{-1} H$_3$BO$_3$, 0.014 g.l^{-1} ZnSO$_4$,7H$_2$O, 0.002 g.l^{-1} FeCl$_3$,6H$_2$O. The medium for xanthan C production contained 0.4 g.l^{-1} NH$_4$Cl, 0.1 g.l^{-1} yeast extract, 10 g.l^{-1} glucose, 2 g.l^{-1} K$_2$HPO$_4$, 0.5 ml.l^{-1} H$_3$PO$_4$ 85%, 0.05 g.l^{-1} Na$_2$SO$_4$, 0.1 g.l^{-1} MgCl$_2$,6H$_2$O, 0.023 g.l^{-1} CaCl$_2$, 0.006 g.l^{-1} H$_3$BO$_3$, 0.007g.l^{-1} ZnCl$_2$, 0.002 g.l^{-1} FeCl$_3$,6H$_2$O.

Methods of analysis. The total protein concentration of crude solutions, including cellular proteins, was determined by the biuret method (*12*). With dilute xanthan solutions, this method was not sensitive enough, and a spectrofluorometric assay was carried out. This method consisted of an acid hydrolysis of the samples followed by derivatization with o-phthaldialdehyde and HPLC analysis on a C18 reverse column of the amino acids released, using spectrofluorescence as the detection mode.

The polymer concentration was determined, either by measuring the dry weight after precipitation of xanthan in an equivolumetric mixture of ethanol and acetone containing KCl, or by sugar determination according to the anthrone method (*12*).

For dilute solutions, a total carbon analyzer (Dorhmann) was used to determine the polysaccharide concentrations. The solutions were obtained after dilution and centrifugation at 50,000 g for one hour and then ultrafiltration on a membrane with a 20,000 molecular weight cut off.

The molecular weight of the polymer was determined by light scattering diffusion on dilute solutions. Pyruvate and acetate contents were estimated by enzymatic methods (*13,14*) after acid hydrolysis of the samples using analyzis kits from Boehringer.

Viscosimetric determinations. The Newtonian intrinsic viscosity of the xanthan molecule was determined by measuring the viscosities of several dilute polymer solutions with a Contraves Low-Shear viscometer. Extrapolation at zero polymer concentration of the reduced specific viscosity gave the value of the intrinsic viscosity, and the Huggins constant was calculated from the slope of the curve.

Concentrated solutions were rheologically characterized by their flow curves, shear stress versus shear rate, with a Rotovisco RV20 Haake viscometer. For highly viscous solutions with characteristics close to those of a gel, assays were carried out at a fixed shear rate once a stable value had been obtained.

RESULTS AND DISCUSSION

Effect of fermentation conditions on xanthan solution properties.

The first step of this study was undertaken in order to correlate the fermentation conditions to the properties in solution of the xanthan produced. Some of the characteristics of xanthan solutions A, B and C obtained in different conditions are presented in Table II. The analyses presented in this Table were carried out on the final samples in the case of batch cultures, i.e. at the moment when the initial glucose was totally consumed. The doublet or triplet data of Table II resulted from repeated fermentations. We have verified that, during the culture in the two batch fermentations, the intrinsic viscosity remained nearly constant.

High concentrations of xanthan were obtained in each case. The intrinsic viscosity was about 5 $m^3 \cdot kg^{-1}$ and the Huggins constant was around the theoretical value of 0.4 for xanthan samples A. These values were higher when the polysaccharide came from a synthetic medium. Furthermore, considering the fermentations on synthetic media, when xanthan was produced in a continuous mode, the intrinsic viscosity was higher and the Huggins constant lower than when it was produced in a batch mode. The pyruvate substitution degree of the macromolecule was dependent on the type of medium used and it was 40% when X. campestris was grown on complex media whereas it was 20% when cultivated on synthetic ones, whatever the mode of culture. Molecular weights were determined in 5 $g.l^{-1}$ NaCl and found to be equal to $7x10^6$ daltons for xanthan A and higher than $10x10^6$ daltons for xanthan C.

The main difference between the mode of cultivations in synthetic and complex media is that, when the microorganism is cultivated in complex media, the growth rates and synthesis rates are higher than those obtained when it is grown on synthetic ones. The values of the intrinsic viscosities suggest that the lower the metabolism rate, the higher the molecular weight and consequently the average length of xanthan chains. Another interesting conclusion is that, considering the values of the Huggins constants and of the pyruvate degrees, even if the composition of the medium is well controlled, especially its protein content, there are other parameters of the fermentation, e.g. sulfur limitation, which can have an action on the chemical structure of the polysaccharide and on its properties in solution. This can explain why the Huggins constant can be higher in synthetic media than in complex one in spite of a lower protein concentration. These modifications of the polymer can be ascribed to a change of metabolism because of different culture conditions. So, both protein presence and also conditions of synthesis seem to play a role of prime importance in the processus of interactions between xanthan chains.

Effects of protein addition on xanthan solution properties.

In order to better understand and confirm the possible role of proteins in xanthan aggregation, the effects of the addition of a protein-rich solution were investigated both in dilute and in concentrated domains of concentrations. In the dilute domain, it should be possible to characterize changes of the molecular size by measuring the intrinsic viscosity whereas concentrated solutions are suitable for estimating aggregation phenomena and can be characterized by their rheological properties (estimated by rheograms).

Addition to purified and diluted solutions of xanthan. For this study, it was necessary to prepare a non-aggregated xanthan solution which was obtained by extensive ultrafiltration of a commercial xanthan sample which was initially non-aggregated. The absence of aggregation was confirmed by the Huggins constant which was 0.4 and the intrinsic viscosity which was 6.7 $m^3 kg^{-1}$. This corresponds to a molecular weight of 4.8×10^6 daltons. This xanthan solution was adjusted at a polymer concentration of 0.4 $g.l^{-1}$ in a protein-rich solution such as corn steep liquor (CSL). Before use, the corn steep solution was centrifuged and only the clear supernatant was added to the xanthan solution. The solvent was 0.1 M sodium chloride and the ratio of protein to xanthan was 10% (w/w).

Intrinsic viscosity determinations (Fig.1) were carried out with xanthan solutions prepared in the presence or not of corn steep liquor. The main results are summarized in Table III.

It can be seen that the addition of corn steep liquor to a xanthan solution led to a drastic decrease in the viscosity in conjunction with an increase of the Huggins constant (Fig.1). This decrease of the macromolecule dimension, due to an entanglement of the polymer chains, and the prevalence of polymer-polymer interactions to solvent interactions, indicate that the polysaccharide molecules are aggregated. A possible explanation of this phenomenon is that proteins from corn steep liquor can induce interactions between xanthan chains, forming xanthan-protein complexes.

It is also worth noting that these interactions were strengthened when temperature was increased since the Huggins constant value rose from 1.1 to 1.5 only after heating during 12 hours at 65°C of the xanthan - corn steep liquor mixture.

Addition to concentrated solutions of xanthan (broth or powder). Rheological measurements were performed on xanthan solutions prepared by dilution of broths, 4800 CX, A, C or powder Rhodopol 23, adjusted at similar polymer concentrations (between 5 and 7 $g.l^{-1}$). The effects of the addition of a protein-rich solution and of other parameters such as salinity and pH were further investigated.

According to fermentation conditions and to downstream processes, the nature and the concentration of proteins in broths may be quite different. Table I indicates the total concentration of proteins, i.e intracellular proteins and proteins used as growth factors during fermentation (CSL, yeast extract). Furthermore, the pyruvate content of the xanthan molecule also depends on fermentation conditions.

The flow curves for concentrated xanthan solutions in 0.1 M sodium chloride are shown in Figure 2. The straight lines obtained from experimental points had a

Table I: Xanthan sample characteristics

Product	Rhodopol 23	4800 CX	A	B	C
Form	powder	broth	broth	broth	broth
polymer g/kg broth g/kg powder	- 640	84 - 74 -	15 -	6 - 9 -	5 -
protein (%) w/w polymer	< 1.4	4.9	23.0	-	12.2
DS Acetate (%)	100	100	100	100	100
DS Pyruvate (%)	30	80	45	20	20

DS = Degree of Substitution

Table II : Characteristics of the different types of xanthan produced

Xanthan type	Fermentation conditions	Intrinsic viscosity $(m^3\ kg^{-1})$	Huggins constant k'	DS Pyruvate (%)
A	batch culture and complex medium	5.0 3.3	0.44 0.45	40 45
B	batch cultutre and synthetic medium	6.1 6.1	1.00 0.70	20 20
C	S - limiting continuous culture and synthetic medium	9.5 7.8 7.2	0.60 0.60 0.40	20 20 20

DS = Degree of Substitution

Table III: Effect of the addition of corn steep liquor (CSL) to a xanthan solution

Diluted solution	Intrinsic viscosity $(m^3\ kg^{-1})$	Huggins constant k'
purified	6.7	0.4
with CSL	4.3	1.1
with CSL and heated	4.3	1.5

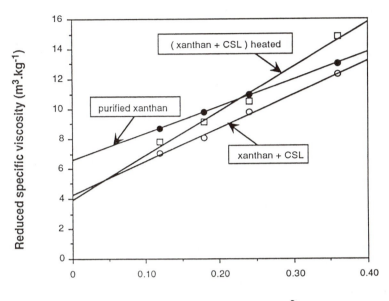

Figure 1: Effect of the addition of corn steep liquor on purified and diluted xanthan solution - Intrinsic viscosity determination in 0.1 M NaCl.

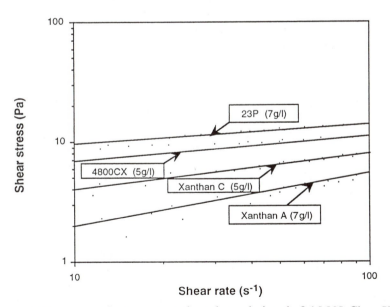

Figure 2: Flow curves for concentrated xanthan solutions in 0.1 M NaCl at pH 7.

slope less than 1.0 for all xanthan solutions (between 0.2 and 0.45) indicating that the solutions had a non-Newtonian behavior in the range of the shear rates investigated. Whatever the fermentation conditions, all the products had the standard behavior of a pseudoplastic fluid where the shear stress increased versus the shear rate according to a power law.

Effect of pH and salinity. Different solutions of xanthan C were adjusted to pH 7 or 8, in 0.1 M NaCl or in reconstituted sea water, in order to study the effect of pH and salinity. The composition of sea water was the following: 27.80 $g.l^{-1}$ NaCl; 9.54 $g.l^{-1}$ $MgCl_2,6H_2O$; 7.73 $g.l^{-1}$ $Na_2SO_4,10H_2O$; 1.62 $g.l^{-1}$ $CaCl_2,2H_2O$; 0.03 $g.l^{-1}$ $SrCl_2,6H_2O$.

For each solution, the variations of the shear stress when the shear rate was increased and then decreased are shown in Figure 3. In sodium chloride salt, whether at pH 7 or 8, the flow curves were identical. In sea water and at pH 7, the same curve was obtained. In sea water and at pH 8, the viscosity of the broth increased, and the variation of shear stress with an increasing shear rate was quite different from the variation observed with a decreasing shear rate.

It is thus clear that in the presence of monovalent cations, xanthan solutions behaved like a pseudoplastic fluid. When some multivalent ions were added, the flow behavior of the solution depended on the pH. At pH 8, the solution behaved like a complex fluid with some thixotropic effect which was not measured at pH 7. According to the pH value, salts such as sodium or calcium can be in different chemical forms. For instance, the amount of hydroxide formed, such as $CaOH^+$ and $MgOH^+$, becomes significant above pH 8. It has been reported in a previous paper that, in the presence of sodium and calcium chloride salts, xanthan macromolecules aggregated and even precipitated, depending on pH, on the mono-bivalent ion ratio, on the pyruvate content and on the purity of the xanthan sample (*15*). This precipitation of xanthan in the presence of bivalent cations, occurring at pH higher than 7, was explained by the reactivity of calcium hydroxide ions on the xanthan chain which was greater than that of bivalent calcium cations (*15*). $CaOH^+$ species are less solvated than Ca^{++} species and then interacted more favorably with the polymer. In the present case, aggregation occurred in sea water, i.e. when calcium and also magnesium ion contents were not negligible in respect to the sodium ion concentration, and at pH 8 where calcium and magnesium hydroxydes begin to appear. Following the same pattern as precipitation, in the present conditions, strong aggregation of xanthan chains may occur and this can explain why the concentrated solution of xanthan C presented some characteristics close to those of a gel and why some thixotropy appeared.

Effect of the addition of proteins. The effect of protein addition to a solution of xanthan C was investigated at pH 8 in 0.1 M NaCl or in sea water. The complex protein solution used was a 75:25 (v:v) mixture of the soluble fractions of corn steep liquor and yeast extract suspensions. It was added to xanthan at a ratio of 50% (w/w). Control samples without protein were also prepared at the same pH. As observed above, xanthan solutions behaved like very viscous fluids at pH 8 and in sea water, so stress measurements were performed at fixed shear rates (50, 100 and

150 s^{-1}) after a stabilization time, which could be one hour in some cases. The experimental points obtained are shown in Figure 4.

In sodium chloride brine, the addition of proteins did not alter the pseudoplastic behavior of xanthan. In sea water, even with a long stabilization time for each measurement, stress values were not identical when the shear rate was increased or decreased, indicating a thixotropic behavior. Protein addition did not significantly enhance this phenomenon. Thus, thixotropy seems to be related rather to the presence of cations inducing an aggregation of xanthan chains than to the addition of proteins. Nevertheless, in sea water, the gel formed by the extra-addition of proteins seems to have a more compact consistency than that of the gel induced by multivalent ions alone. This was probably the result of a synergic effect, e.g. the expulsion of some water molecules from the interior of the gel network by proteins. Experiments such as the determination of the elasticity module could bring some complementary information about the structure of the gel formed in these two cases.

Effect of the presence of cells. As xanthan C was used as a crude broth, it contained cells of *X. campestris.* A cell-free solution was prepared by centrifugation at 50,000 g during 11 hours. The effect of the addition of the complex protein mixture to sea water at pH 8 was studied as described in the previous section. Figure 5 gives the flow curves of xanthan C with and without cells.

As previously observed, the xanthan C solution in sea water in the presence of proteins behaved like a thixotropic fluid. By contrast, an almost standard rheological behavior of a pseudoplastic fluid was found with the centrifuged solution in sea water with proteins added. A weak thixotropic effect was still observed probably because of the difficulty to completely eliminate cells from such a viscous fluid.

As described above, the formation of xanthan-protein complexes induced by salts such as calcium or magnesium chloride depends strongly on the quality of the solution. It was previously reported that when xanthan solutions contained less impurities, interactions between macromolecules and calcium ions were less important, and that the pH limit corresponding to precipitation was shifted to basic pH (*15*). Here, when xanthan C broth was centrifuged, thixotropic phenomena were less apparent. So, for this xanthan solution, it is clearly shown that the presence of cellular components was essential to induce interactions between xanthan chains and cations. It can also be noted that the xanthan broth was frozen before all the experiments and thus, that the cells may have been lysed thereby releasing all their constituents inside the medium. Because of the diversity and complexity of the cellular components, it is difficult to predict which kind of interactions can be involved in the aggregation.

Effect of the origin of xanthan samples. Different representative samples of xanthan were tested for their ability to display thixotropic behavior in the suitable conditions determined from the aforementioned results. Following the same procedure, all xanthan samples were dissolved in sea water at pH 8 in the presence of added proteins. The flow curves, i.e stabilized values of shear stress at three shear rates, were determined and are shown in Figure 6. Under these conditions, where the aggregation of xanthan chains was favoured, only xanthan C exhibited thixotropic behavior. The other products had the standard behavior of a pseudoplastic fluid and the shear stress values were identical when the shear rate was increased or decreased.

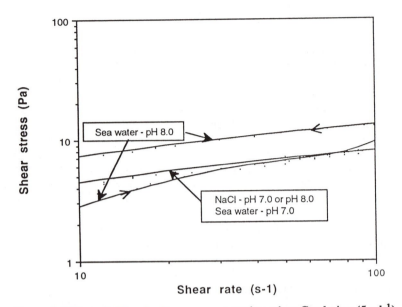

Figure 3: Effect of pH and salt on concentrated xanthan C solution (5 g.l^{-1}).

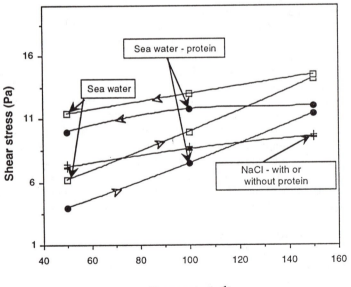

Figure 4: Effect of added proteins on concentrated xanthan C solution (5 g.l^{-1}) in sea water at pH 8.

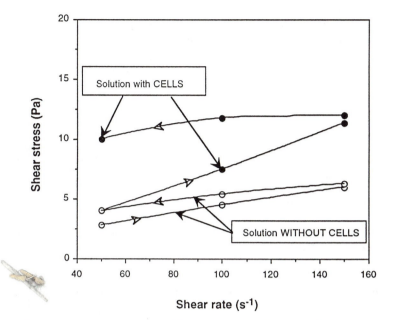

Figure 5: Effect of the presence of cells on concentrated xanthan C solution (5 g.l^{-1}) in sea water at pH 8 with added proteins.

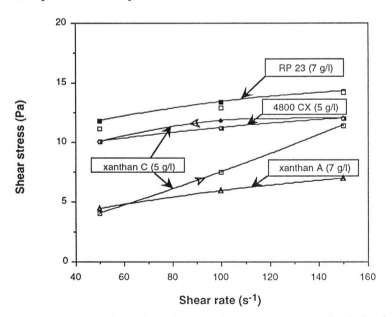

Figure 6: Effect of the origin of xanthan samples on a concentrated solution in sea water at pH 8 with added proteins.

The main differences between xanthan C and the other products were the degree of pyruvate substitution of xanthan, the nature of the fermentation medium, the protein concentration and the intrinsic viscosity (see Table II).

The higher the ionization degree of the xanthan molecule, the stronger the interactions between chains and ions as calcium. In the present case, the ionization degree of xanthans was dependent only on their pyruvate content since glucuronic acid and acetate contents of the different samples were identical. Xanthan C had the the lowest pyruvate content and consequently the smallest ionization degree of all the samples. Thus, its special behavior cannot be explained by this factor.

Another difference was the protein quantity which is related to cell concentration. The attempts to determine the amount of cells of samples by dry weight or centrifugation succeeded only with xanthan C. This result indicates that the cellular concentration was higher in this particular broth or that the cells were completely lysed in the other xanthan samples.

Moreover, xanthan C was issued from a sulfur-limited continous culture on a synthetic medium, whereas the other ones were certainly produced in a batch mode in complex media. It is known that a sulfur limitation modifies the composition of the cell walls or membranes of *X. campestris* (6).

Finally, as the intrinsic viscosity of xanthan C was higher, its molecular weight can also be higher. It is quite possible that interactions between xanthan and bivalent ions and proteins can be different according to the polysaccharide chain length.

So, the aggregation of xanthan molecules was induced by the presence of both cations and proteins. Furthermore, the aggregation of xanthan C chains was probably enhanced by the particular nature of the fermentation conditions in addition to the interactions with calcium or magnesium hydroxide ions and added proteins. These peculiar conditions of culture seem to be indirectly involved in the mechanism of aggregation by the cells themselves and possibly by the chain length of the xanthan synthesized.

CONCLUSION

The rheological behavior of xanthan samples obtained in various fermentation conditions with dilute or concentrated polymer solutions in the presence of proteins has been investigated.

- The characteristics of the xanthan solutions obtained are greatly dependent on fermentation conditions.

- The addition of proteins can induce an aggregation of xanthan chains according to the ionic composition of the solvent (presence of cations such as calcium or magnesium) and to the pH. Furthermore, the flow behavior of xanthan-protein solutions exhibited some thixotropic phenomena

-The cells, the nature of which depends on fermentation conditions, had a particular effect on the interactions between xanthan chains, cations and added proteins.

ACKNOWLEDGMENTS: The authors would like to thank E. Praet, S. Prigent, Y. Benoit and Y. Le Penru for excellent technical assistance.

REFERENCES

1. Audibert, A., Lecourtier, J., and Lund, T., VIth Eur. Symp. Improved Oil Recovery, Proc., Stavanger, Norway, May 21-23, 1991.
2. Darley, H.C.H., and Gray, G.R., in "Composition and properties of drilling and completion fluids" 5th ed, Gulf Publishing Company, Book Division, Houston, London, Paris, Tokyo, 1988.
3. Hassler, R.A., and Doherty, D. H., *Biotechnol. Progr.* **6**, 182-187, (1990).
4. Smith, I.H., Symes,K.C., Lawson, C.J., and Morris, E.R., *Int. J. Biol. Macromol.* **3**, 129-134 (1981).
5. Tait, M.I., Sutherland, I.W., and Clarke-Sturman, A.J., *J. Gen. Microbiol.* **132**, 1483-1492 (1986).
6. Evans, C.G.T., Yeo, R.G., and Ellwood, D.C., *in* "Microbial Polysaccharides and Polysaccharases" (R.C.W. Berkeley, G.W Gooday and D.C. Ellwood Eds.), p.51-68, Academic Press, London, New York and San Francisco, 1979.
7. Allain, C., Lecourtier, J., and Chauveteau, G., *Rheol. Acta* **27**, 255-262 (1988).
8. Kolodziej, E.J., 62nd Annual Tech. Conf. Exh. Soc. Pet. Eng. SPE 16730, Dallas, Sept. 27-30, 1987.
9. Monot, F. , Benoit, Y., and Ballerini, D., 190th ACS Ntl. Meeting, Chicago, Sept., 8-13, 1985.
10. Jousset, F., Green, D.W., Willhite, G.P., and McCool, C.S., VIIth Symp. Enhanced Oil Recovery, Tulsa, Oklahoma, April 22-25, 1990.
11. Morris, V.J., Franklin, D., and I'Anson, K., *Carbohydr. Res.*, **121**,13-30 (1983).
12. Herbert, D., Phipps, P.J., Strange, R.E., *in* "Methods in Microbiology", (J.R. Norris and D.W. Ribbons, Eds.), vol 5B, p.209, Academic Press, London 1971.
13. Czok, R., and Lamprecht, W., *in* "Methods of Enzymatic Analysis", 2nd ed, (H.U. Bergmeyer Ed.), 3, p.1446-1451, Verlag Chemie, Weinheim /Academic Press, Inc., New York and London, 1974.
14. Beutler, H.O., *in* "Methods of Enzymatic Analysis", 3rd ed., (H.U. Bergmeyer Ed.), 6, p.639-645, Verlag Chemie, Weinheim, Deerfield Beach/Florida, Basel, 1984.
15. Lecourtier, J., Noïk, C., Barbey, P., and Chauveteau, G., Proc. 4th Eur. Symp. Enhanced Oil Recovery, Hamburg, Germany, Oct. 27-29, Oct. 1987.

RECEIVED January 27, 1993

Chapter 20

Measurement of Steric Exclusion Forces with the Atomic Force Microscope

A. S. Lea, J. D. Andrade, and V. Hlady

Department of Bioengineering, University of Utah, Salt Lake City, UT 84112

Atomic force microscope probes were modified by attaching polyethylene oxide to the silicon nitride cantilever tip. Two high molecular weight species, 200 and 900 kDa, were attached by physical adsorption from a dilute polymer solution. One low molecular weight species, 2 kDa, was chemically bound to the surface. Force-distance plots were obtained for modified tips and freshly cleaved mica and for plasma cleaned, unmodified tips and PEO adsorbed silicon nitride substrate in the presence of 0.1 M KNO_3. force-distance plots were also obtained in a 0.1 % w/v aqueous polyethylene oxide (M_w 900 kDa) solution containing 0.1 M KNO_3 with tips that were plasma cleaned only. The force-distance plots show the existence of a steric exclusion force as the tip and sample are brought closer together, when the high molecular weight polyethylene oxide is adsorbed on either the substrate or the tip and when plasma cleaned, unmodified tips, incubated for at least 8 hr in the polyethylene oxide solution, are used.

The operation of the atomic force microscope (AFM) relies upon the intermolecular forces that are exerted on the probe tip by the surface of a closely placed sample (1-3). The majority of the AFM research has relied heavily upon hard core repulsion forces, otherwise known as 'contact forces', to produce the images that are widespread in the literature. Naturally, this is the intermolecular force of choice when obtaining atomic scale images, since it is the only one capable of providing such high resolution using the currently available probe tips (4). When operating in liquid environments, it is no longer the hard core repulsion, but the hydration forces that exist between the probe tip and sample that produce these images (5-6). These forces are short-range forces and are predominant at tip-sample separations less than a few angstroms.

Because of the macroscopic nature of the tip and the sample, long-range intermolecular forces also become important, especially at distances greater than a few angstroms. Long-range intermolecular forces that have been utilized to date with the AFM are attractive van der Waals forces (2,7), electrostatic forces (6,8,9), and magnetic forces (10-11). This paper describes the use of long-range steric exclusion repulsive forces in atomic force microscopy. This was accomplished by attaching water soluble polyethylene oxide (PEO) to the probe tip and by operating the instrument with the probe in an aqueous PEO solution.

0097–6156/93/0532–0266$06.00/0

In good solvents, surfaces covered with polymer chains repel each other when brought together due to the steric exclusion forces manifested by the extended polymer chains (*12-14*). The origin of these repulsive forces is attributed to two components: an elastic component and an osmotic component (*15-16*). The elastic component arises from the chain segments that have a tendency to extend themselves upon compression. The osmotic component arises from the local increase in chain segment concentration upon compression resulting in a loss of configurational entropy.

The onset of the steric exclusion force depends on the means of attaching the polymer chains to the substrates. For physically adsorbed polymer chains covering both surfaces, the steric exclusion force becomes detectable around $6R_g$, where R_g is the unperturbed radius of gyration of the random polymer coil in solution (*12*). For terminally attached polymer chains, the repulsion commences around $12R_g$ (*14*). These values are approximate and depend on a number of factors including solvent quality, temperature, surface concentration and type of polymer chains attached to the surfaces.

If the steric exclusion force commences at a larger separation distance than the attractive van der Waals force, then the steric exclusion force dominates the attractive van der Waals force and a monotonically increasing repulsion is observed in the force-separation distance profiles. This occurs when attaching polymer chains containing a large number of segments, i.e. at large M_w, to the surface. Attaching low molecular weight polymer chains shifts the onset of the steric exclusion force closer to the surface and the attractive van der Waals component could dominate, resulting in an overall attractive interaction in a region of the force-distance profile.

We have modified conventional AFM tips by attaching PEO of different molecular weights to the tips with the intention of using steric exclusion forces as the predominate imaging force. PEO with molecular weight of 200 kDa or 900 kDa was physically adsorbed to the cantilever tips, whereas low molecular weight PEO (or PEG: polyethylene glycol) was covalently bound to the tip to prevent desorption of the polymer chains. It is possible that these cantilevers could greatly reduce the lateral translation of surface adsorbed entities in aqueous solutions (*17-18*). The AFM tips covered with polymeric PEO chains are expected to provide a more forgiving imaging force.

Experimental

Materials. Gold-chromium coated silicon nitride cantilevers with integrated tips (spring constants of 0.064 N/m) were obtained from Park Scientific Inc. PEO with molecular weights of 200 kDa and 900 kDa, polyethylene glycol monomethyl ether (PEG) with a molecular weight of 2 kDa, and 3-aminopropyltriethoxysilane were obtained from Aldrich Chemical Co.. The water used in this study was microelectronics lab quality deionized water, which was subsequently passed through an organic removal cartridge and a particle filter. Muscovite mica was obtained from Asheville-Schoomaker. Silicon wafers with a 250-300 Å CVD silicon nitride coating were obtained from Hedco lab at the University of Utah. The AFM is a NanoScope II from Digital Instruments.

Methods. Physical adsorption. PEO was adsorbed either onto the silicon nitride cantilever tip or onto a piece of silicon nitride coated silicon wafer. The substrates were cleaned by placing them in an oxygen plasma (200 mm Hg, 25 W) for 5 minutes. Adsorption of the PEO to the clean surface was accomplished by placing the cantilever or the wafer piece in a 0.1% w/v solution of PEO in water. Adsorption was allowed to take place for at least 18 hours (sometimes as much as a week). The substrate was then placed in water for 22-24 hours to allow desorption of the weakly bound polymer chains. The substrate was vacuum dried prior to use.

Adsorption of PEO was also accomplished *in situ* by mounting a plasma cleaned cantilever to the AFM fluid cell. The fluid cell was mounted on a freshly cleaved mica surface. A 0.1% w/v solution of PEO in 0.1M KNO_3 was injected into the fluid cell.

Chemical binding. Synthesis of the aldehyde-terminated PEG (PEG-CHO) was accomplished by a modification of the acetic anhydride method of Harris et al.[19] All solvents, excluding acetic anhydride, were dried over molecular sieves. The formation of aldehydes was monitored at 560 nm using the Schiff reagent [19].

The silicon nitride cantilevers with tip were treated with an oxygen plasma to remove carbon contamination from the surface. Amine groups were incorporated onto the surface by placing the cantilever in a 5% v/v 3-aminopropyltriethoxysilane (APS) in water solution for 10 minutes and then rinsing thoroughly with water [20].

Chemical binding of the PEG 2 kDa to the APS derivatized cantilevers was achieved by placing the cantilevers into a 50 mg/ml solution of PEG-CHO in a pH 5.2 acetate buffered 11% w/v K_2SO_4 solution at 60°C for 40 hours [20]. The cantilevers were rinsed with water and placed in a 8.45 mg/ml solution of $NaCNBH_4$ in water at room temperature for 4 hr to reduce the Schiff base. The cantilevers were rinsed with water and vacuum dried before use.

Force plots. The cantilever tips were positioned near a freshly cleaved mica surface in the AFM fluid cell. 0.1 M KNO_3 was injected into the fluid cell and AFM force-distance plots were obtained by oscillating the piezoelectric crystal with the sample in the z-direction. This commercial instrument requires that the tip 'engages' with the surface before any imaging or force plots are obtained. Except for a few cases, all force plots were obtained from the commercial AFM software. The other method for obtaining force-distance plots uses an external data collection system. The photodiode signal, [A-B]/[A+B], and the voltage signal that drives the piezoelectric crystal in the z direction were collected by a HP 24000A digital oscilloscope (7-bit resolution), which could average the waveforms and store them in memory. A computer program was written to collect and plot the data from the oscilloscope using an HP 9300 computer.

Ellipsometry. Measurements of the index of refraction and film thickness were performed on a Rudolph Research model 43603-200E ellipsometer with a 642.8 nm He-Ne laser.

Computer Modelling. The computer program POLAD [21] was employed to model the distribution of trains, loops, and tails of the PEO physically adsorbed to the silicon nitride surface. This program is based on interaction parameters of the polymer segments with the surface and the solvent (c_S and c) and considers the chains as connected sequences of segments. For the physically adsorbed PEO, the interaction parameters were 1.00 for the surface and 0.45 for the solvent. The bulk volume fraction, f_b, was 10^{-3}. Milner has made a quantitative comparison of the experimental forces generated by compression of terminally attached polystyrene brushes with those calculated from self-consistent field equations [22]. The theoretical forces are in good agreement with those obtained experimentally, provided the chains are not compressed too much. The program GOLIAD [23] was used here to model the distribution of terminally attached PEG segments and to model the free energy change upon chain compression. A covalent bond was simulated by increasing the interaction energy of one terminal segment until a stable attachment occurred.

Results and Discussion

ESCA studies of the 'as received' cantilever tips (*24*) show a considerable amount of carbon on the surface (27%) (Table 1). A high resolution scan of the C(1s) peak shows 82% of the carbon is aliphatic. Oxygen plasma treatment of the cantilever tips shows a decrease in the carbon signal from 27% to 5% indicating that plasma treatment removes this contamination layer. Oxygen plasma treatment also enriches the surface concentration of oxygen from 19% to 33%. We opted for this treatment as chromic acid etching would remove the gold-chromium backing on the cantilever necessary for reflection of the laser beam. The outer 6 - 9 Å of the silicon nitride surface has been reported to be a silicon oxynitride of variable stoichiometry (*25-27*). The outermost layer of the surface is essentially silicon oxide and behaves just like a silica surface electrophoretically (*28*). Oxygen plasma treatment then would introduce enough silanol groups to the surface for silane chemistry to occur.

The covalent binding of PEG-CHO 2 kDa on silicon nitride coated silicon wafer pieces was monitored by ESCA. The surface of these wafer pieces were shown by ESCA to be very similar to that of the cantilever tips, especially following oxygen plasma treatment (ESCA results of the oxygen plasma treated wafer piece are also shown in Table 1). The APS derivatized cantilevers show an increase in the carbon signal (from 2 to 19%), a slight reduction in silicon signal, and a reduction of the oxygen signal. In addition, there is a slight increase in the binding energy bandwidth of the N(1s) peak. These results are consistent with the binding of APS to the surface of the silicon nitride coated wafer (*20*). The PEG bound surfaces show a decrease in both carbon and oxygen content, which is not expected. Yet, the FWHM bandwidth of the O(1s) peak has increased from 1.71 and 1.75 eV for the oxygen plasma treated and APS bound surfaces to 2.30 eV for the PEG bound surface, indicating the presence of an additional chemical species of oxygen. A peak-fitting routine of a high resolution oxygen spectrum reveals an additional peak shifted to lower binding energy by 1.2 eV that comprises 10% of the total oxygen signal. Furthermore, a high resolution spectrum of the C(1s) peak shows that 72% of the carbon signal is in the ether form. These last two results indicate the presence of PEG on the surface of the wafer piece.

Sample	Silicon	Oxygen	Nitrogen	Carbon
As received cantilever	31 %	19 %	22 %	27 %
Plasma treated cantilever	39 %	33 %	22 %	5 %
Plasma treated silicon nitride coated wafer	36 %	20 %	39 %	2 %
APS bound wafer	33 %	27 %	20 %	19 %
PEG 2 kDa bound wafer	36 %	12 %	36 %	16 % (72 %)
PEO 900 kDa adsorbed wafer	34 %	30 %	23 %	12 % (51 %)

Table 1. Summary of ESCA data for silicon nitride cantilevers 'as received' and oxygen plasma modified and for a silicon nitride coated silicon wafer 'as received', oxygen plasma modified, APS modified, PEG bound and PEO adsorbed. The values are in atomic percent. The numbers in the parentheses in the carbon column indicate what percentage of carbon is in the ether form

For the physically adsorbed PEO, high molecular weight polymer was be used to prevent spontaneous desorption of the polymer. Although each segment of the PEO adsorbed to the surface may have an interaction energy of a few tenths of a kT, there are so many contacts that adsorption is essentially irreversible (29). Any loosely bound polymer was allowed to desorb by placing the cantilever or silicon nitride coated silicon wafer in polymer-free solution. The ESCA results indicate that of the 12 % of the carbon detected on the surface, 51 % of it is in the ether form. This demonstrates that PEO is present on the surface.

Ellipsometric measurements of the PEO 900 kDa physically adsorbed layer on the plasma cleaned silicon nitride coated wafer indicate that the PEO layer thickness in air is 1.94 ± 0.10 nm using a measured refractive index of 1.94 for silicon nitride and an assumed refractive index of 1.45 for PEO in air.

The results of the computer modelling are shown in Table 2. The POLAD program produces the total adsorbed amount in equivalent monolayers, the ellipsometric thickness in number of segments, and hydrodynamic thickness in number of segments. Thicknesses in nm are calculated using 0.29 nm as the PEO segment length (30). The ellipsometric thickness of the physisorbed PEO 900 kDa layer was calculated to be 1.98 nm. This agrees with the experimental measurement of 1.94 ± 0.10 nm. Also shown in the table is the total amount adsorbed and hydrodynamic thickness. Although, the surface has similar adsorbed amounts, 2.13 vs. 1.98 equivalent monolayers, the hydrodynamic thickness is substantially different. The PEO 900 kDa layer has a hydrodynamic thickness 6 times larger than the PEO 200 kDa layer, at identical hydrodynamic permeabilities. This large difference is attributed to the extension of the tails into the solvent (31). Although the hydrodynamic permeability is not known, it varies over a range of 0.5 to 2.0. Using these values, the hydrodynamic thickness varies between 64 and 102 nm for the physisorbed PEO 900 kDa chains and between 9.9 and 16 nm for the physisorbed PEO 200 kDa chains.

Table 2. Summary of POLAD modelling results for physically adsorbed PEO (900 kDa and 200 kDa) on silicon nitride using a hexagonal lattice, c=0.45, c_S=1.0, and a volume fraction of 10^{-3}

System	Adsorbed amount (equivalent monolayers)	Ellipsometric. thickness (nm)	Hydrodynamic. thickness (nm), (c = hydrodynamic permeability (21))
adsorbed PEO 900 kDa	2.13	1.98	102 (c=0.5) 84 (c=1) 64 (c=2)
adsorbed PEO 200 kDa	1.98	1.73	16 (c=0.5) 13 (c=1) 9.9 (c=2)

Y. S. Lin et al. have measured ellipsometrically the maximum amount of PEG 2 kDa bound to a silica surface using the same chemistry described in this manuscript (20). The amount corresponded to 1.8 equivalent monolayers or one PEG 2 kDa chain per 5 nm^2. Using GOLIAD we calculated the free energy change for compression of two flat surfaces of PEG 2 kDa chains terminally attached carrying one equivalent monolayer (Figure 1). For one surface, this free energy change is approximately 1 kT per site $(0.3 \text{ nm})^2$ when the separation distance reaches 0.6 nm (2 layers) corresponding to a free energy change of 4.6 x 10^{-18} J nm^{-2} for one surface. Using the Derjaguin approximation relating the force between a sphere and a flat surface to the energy between two flat surfaces, $F(D) = 2\pi RW(D)$, and a radius of curvature of 20 nm for the silicon nitride tip (*32*), a force of 5.7 nN should be observed at a tip surface separation distance of 0.6 nm. At this separation we have calculated a non-retarded van der Waals attractive force of 0.5 nN using a Hamaker constant of 0.5 x 10^{-19} J. The AFM is capable of measuring such forces and it is expected that steric exclusion forces would appear in the force-distance curves.

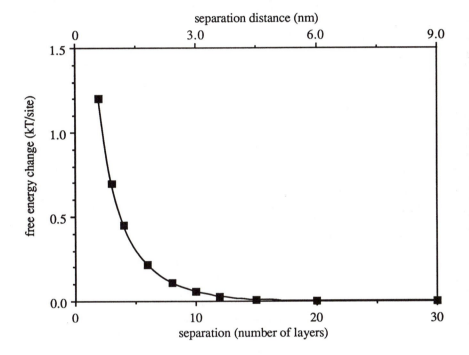

Figure 1. Free energy change due to compression of terminally bound PEG 2 kDa in kT per site as a function of number of layers and separation distance calculated by GOLIAD for 2 flat surfaces.

The AFM force-distance plots were obtained by oscillating the sample up and down and monitoring the response of the AFM cantilever. The sample had to come into contact with the AFM tip and retract from the tip with each oscillation. Under conditions of small cantilever bending, the force exerted on the cantilever is directly related to the extent of bending by $F = k\Delta x$, where Δx is the distance the tip of the cantilever has moved and k is the spring constant of the cantilever.

The AFM force-distance plots are similar, but not identical to the force-distance profiles obtained by the surface force apparatus. The bending of the cantilever can follow the force exerted on it by the sample as long as the spring constant of the cantilever exceeds the force gradient of the exerted force. Otherwise, the cantilever jumps into contact with the surface in an attractive regime and in the repulsive regime, is merely pushed a distance equal to the distance the sample is moved, thus maintaining a constant separation distance between the two. The same is true in the steric repulsion case; once the force gradient exceeds the cantilever spring constant, the polymer layer can not be compressed by the restoring force of the cantilever. This is depicted as a linear portion in the AFM force-distance plot . With the surface force apparatus, the spring is significantly stiffer so that the spring constant exceeds the force gradients of the force-distance profiles. Only at very small separation distances does the force gradient exceed the spring constant of the cantilever. In addition, there is no independent means of determining the absolute separation distance between the tip and the sample with the AFM. The location of the 'zero' separation distance is arbitrarily defined by the position of the sample where the linear portion of the force-distance plot commences. Retraction of the sample from this location leads to larger separation distances.

Figure 2a depicts a typical force-distance plot in the absence of PEO derivatization. The abcissa is the response of the photodiode and is directly related to cantilever bending. The ordinate is the distance the sample has moved when oscillating up and down (larger separations are to the right). At large separation

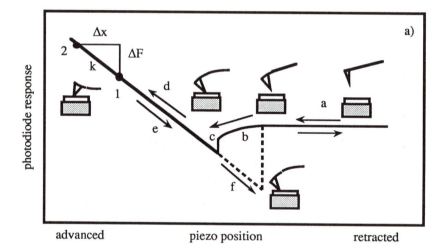

Figure 2a. A schematic of a typical AFM force-distance plot using unmodified tips. The arrows show how the force plot is generated as the sample is advanced and retracted. At (a) the tip and sample are far apart. When the tip gets close enough to experience the attractive van der Waals force, the cantilever starts to bend (b). When the force gradient exceeds the cantilever spring constant, the tip jumps into contact with the surface (c). Once in contact, the tip and sample move the same amount as shown by the linear portion (d). Upon the retraction (e), the cantilever relaxes a distance equal to the amount the sample has been retracted. If there is an adhesive force, then there is hysteresis in the loop (f). The inset depict the state of cantilever bending. Positions 1 and 2 show how to calibrate the force scale. The distance the cantilever has moved, Δx, multiplied by the spring constant, k, yields the force difference, ΔF, between the two positions.

distances (a), there is negligible force exerted on the cantilever by the sample and there is no cantilever bending as indicated by the flatness of this portion of the plot. As the sample is raised up (b), an attractive van der Waals force is exerted on the tip and the cantilever begins to bend downward. At point (c), the force gradient exceeds the spring constant of the cantilever and the cantilever 'jumps' into contact with the surface. Further raising of the sample causes, the cantilever to deflect an equal amount (d). The beginning of this linear portion is defined as 'zero' separation distance. In reality, the cantilever tip can not get closer to the surface than a few Ångstroms. Upon retraction (e), the cantilever relaxes an amount equal to the sample retraction. If there is any adhesion that exists between the tip and sample, the sample must be retracted further for the two surfaces to separate (f).With PEO modified tips (Figure 2b), the steric exclusion force dominates the attractive van der Waals force and a monotonically increasing repulsive force is observed. Only at small separations, where the repulsive force gradient exceeds the cantilever spring constant does the curve become linear, as in the case of an untreated tip. Upon retraction, there is no adhesion and the retraction curve follows the advancing curve.

The abcissa of the plots is already calibrated by the instrument, only the 'zero' separation distance needs to be determined. The ordinate can be easily calibrated to show force units because in the linear region of the force-distance plot, the slope equals k, the spring constant of the cantilever: $\Delta F = k\Delta x$ (see Figure 2a). Zero force is defined by the flat part of the plot.

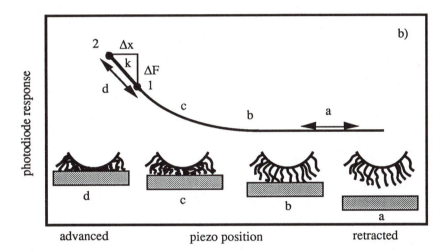

Figure 2b. A schematic of a typical AFM force-distance plot using a PEO modified tip. At (a), there is no interaction between tip and surface. As the chains begin to compress (b) a repulsive steric exclusion force is observed. At (c), the chains are compressed even more producing an even larger repulsive force that dominates the attractive van der Waals force. At (d), the chains are so much compressed that the cantilever spring constant is much weaker than the 'spring constant' of the PEO chains and the cantilever continues to bend upward the same amount as the sample has been moved due to the large repulsive force gradient. Upon retraction, no adhesion is observed (provided there is no bridging) and the curve coincides with the approach curve.

A typical AFM force-distance plot using a plasma cleaned silicon nitride tip and a freshly cleaved mica surface in the presence of 0.1 M KNO_3 is shown in Figure 3 (plot A). The advancing curve and the receding curve should be coincident along the linear and flat portions, but the instrumentation does not always depict this superposition. In this plot, as is the case in all other plots, the advancing curve is shown displaced vertically above the receding curve for clarity. This plot shows the jump into contact upon approach of the sample due to the gradient of the attractive van der Waals force exceeding the cantilever spring constant. There is an adhesive force between the tip and sample upon retraction. This adhesion is presumably due to some contamination that exists on the surfaces (33). The magnitude of this adhesive force is on the order of 0.3 nN (1.0 div • 4.03 nm/div • 0.064 nN/nm). This plot is similar to those obtained by Weisenhorn et al. (3).

Plot B in Figure 3 shows the AFM force-distance curves obtained when a PEG 2 kDa covalently bound to the tip compresses against a freshly cleaved mica sample in the presence of 0.1 M KNO3. A small attractive force just prior to the linear portion of the force-distance plot is observed. This implies that the steric exclusion force at this position is weaker than the van der Waals attractive force. This is not unexpected, since such a small molecular weight PEG is bound to the surface. With a 2 kDa PEG polymer attached, the R_g is 1.7 nm and the steric exclusion force should commence at about 4 nm (2.5 R_g). Computer modelling by GOLIAD indicated that a force of 5.2 nN (5.7 minus 0.5 nN) can be expected at a 0.6 nm separation distance when one equivalent monolayer of PEG segments was present on the surface of the probe. Such a force is detectable by the AFM. One possibility why a repulsive force does not appear in this experiment is that a lesser amount of PEG 2 kDa than one equivalent monolayer is bound to the surface.

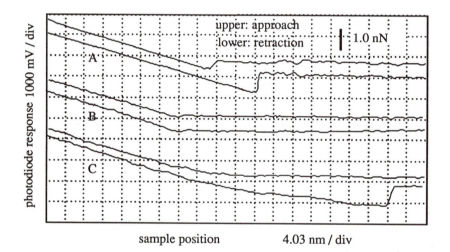

Figure 3. Plot A depicts force-distance curves recorded using a plasma cleaned tip and freshly cleaved mica in the presence of 0.1 N KNO3. These curves display a jump into contact upon approach and an adhesive force upon retraction of the tip. Plot B shows force-distance curves recorded using a PEG 2 kDa treated tip and freshly cleaved mica in the presence of 0.1 N KNO3. This plot does not show a monotonically increasing repulsive force upon approach, but does show an adhesive force upon retraction and is similar to the force-distance plot of a tip plasma cleaned only. Plot C shows force-distance curves recorded using a PEO 900 kDa treated tip and freshly cleaved mica in the presence of 0.1 N KNO3. This plot shows no jump into contact upon approach and an adhesive force upon retraction. The cantilever spring constant was 0.064 nN/nm and the frequency of oscillation was 25 Hz in all cases.

Plot C in Figure 3 shows AFM force-distance curves recorded using a tip coated with physically adsorbed 900 kDa PEO and freshly cleaved mica in 0.1 M KNO₃. Upon approach of the sample to the tip, there is a monotonically increasing repulsive force that commences at about 25 nm from the linear portion of the force curve which signals the onset of the steric exclusion force. This distance is considerably less than the 2.5 - 3.5 R_g observed in surface force apparatus experiments (*12-13*). Since in the AFM experiment the sample and the PEO coated tip have to engage with a moderate force prior to obtaining the force curves, it is likely that some of the PEO has been extruded from the contact area. This would result in a steric exclusion force commencing at smaller separation distances. Also shown in Figure 3, plot C is a 1 nN adhesive force upon retraction. The presence of this adhesive force can be explained by the bridging that occurs between the PEO and the uncovered mica surface (*34*). This is a reasonable explanation since the surfaces were compressed together with a force of 50 - 100 nN. When one considers that the contact area is small, perhaps on the order of 100 nm², the pressure between the silicon nitride tip and the mica is in the MPa to GPa range. This pressure can force the PEO to make contacts with the surface even though the duration of contact is short (25 Hz oscillation of the piezoelectric crystal). One notes that the adhesive force is also non-linear which indicates that there has been a transfer of chains from the tip to the surface which become entangled upon compression and become untangled upon retraction.

Force-distance curves have also been obtained with a plasma cleaned probe and mica in a 0.1% PEO 900 kDa solution containing 0.1 M KNO₃ so that PEO adsorbs on both surface and AFM tip. The first force-distance curve was taken within a few minutes of injection of the PEO solution (Figure 4, plot A). After 8 hr of incubation (Figure 4, plot B), a repulsive force develops for both approaching and receding curves. This phenomenon has also been observed by Klein and Luckham with the surface force apparatus (*34*). The repulsive force commences at a distance of 100 nm from the contact point. This distance is about 6 R_g and agrees well with the observations of Klein and Luckham. Continued compression cycles strip the PEO from the contact area as evidenced by the appearance of an adhesive force upon retraction (Figure 4, plot C). The repulsive force commencing at only 50 nm is another indicator of this PEO expulsion.

To verify the existence of the effect of tip curvature, we have performed identical experiments, but now with the PEO adsorbed only to the silicon nitride substrate. PEO 200 kDa and 900 kDa were physisorbed on a silicon nitride surface and obtained force-distance curves with a plasma cleaned probe. Figure 5 (plot A) shows the initial force-distance curves with PEO 900 kDa. The repulsive steric exclusion force commences at around 50 nm, which is agrees with a 3 R_g distance of 51 nm expected for one modified surface. Again, continued compression cycles results in the development of an adhesive force due to polymer expulsion. Occasionally, force-distance curves similar to that shown in Figure 5 (plot B) will appear. The sawtooth appearances upon retraction are a result of either PEO bridging that occurs between the probe and the surface or chain entanglement of the PEO on both surfaces. If there is bridging, then as the sample is retracted there is resistance to separation due to the bridging. The adhesion, however, terminates sharply when the retraction force 'snaps-off' the PEO chains from the surface. This particular curve shows a series of three successive releases of the PEO from the neighboring surface, each one having longer chain lengths between the bridging points. If there is chain entanglement, however, then the resistance to separation is due to interlocking of the chains from the two surfaces. The three abrupt jumps to zero force would then be a result of the chains becoming untangled or the chains being pulled off the surface by the retracting entanglement.

Figure 6 shows two force-distance plots for a PEO 200 kDa adsorbed silicon nitride surface. The initial curve recorded at the pristine site (Figure 6, plot A) shows the repulsive steric exclusion force upon approach, but an adhesion due to PEO expulsion upon withdrawal. The repulsive steric repulsion force commences at about 25 nm which is in agreement with the 25 nm expected for a 3 R_g distance with only one surface modified. Subsequent compressions at the same site (Figure 6, plot B) show the development of a jump into contact and a large retractive adhesive force, both indications of further PEO expulsion.

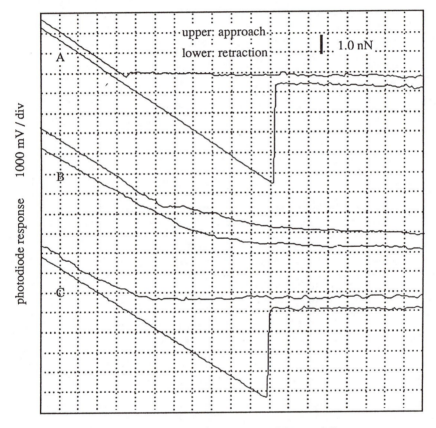

Figure 4. This series of force-distance plots shows the measured forces when a plasma cleaned tip and mica are used to obtain force-distance plots in the presence of a 0.1 % w/v PEO 900 kDa solution containing 0.1N KNO3 as a function of incubation time. Immediately after injection (plot A), following an 8 hour incubation (plot B), and after 2 minutes of oscillations following an 8 hour incubation (plot C). The cantilever spring constant was 0.064 nN/nm and the frequency of oscillation was 1 Hz.

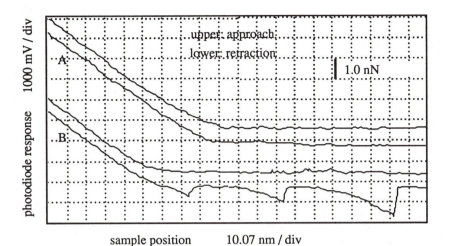

sample position   10.07 nm / div

Figure 5. Two force-distance plots obtained when the PEO 900 kDa has been physically adsorbed to the silicon nitride surface in the presence of 0.1 N KNO3. A pristine site compression cycle is shown by the curves A. Occasionally, a compression cycle would yield a force-plot similar to the curves B. Here, a series of adhesive "snap-offs" upon retraction is most likely due to the ripping of PEO chains off from one of two surfaces after they have bridged the two surfaces. The cantilever spring constant was 0.064 nN/nm and the frequency of oscillation was 1 Hz.

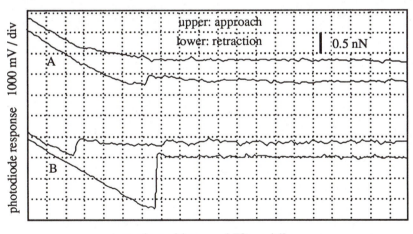

sample position   4.70 nm / div

Figure 6. Two force-distance plots obtained when the PEO 200 kDa has been physically adsorbed to the silicon nitride surface in the presence of 0.1 N KNO3. As in figure 5, an pristine site is used to obtain a force plot (plot A). After several compressions (plot B) a jump into contact and an adhesive component developed indicating that the PEO has been expelled from the contact area. The cantilever spring constant was 0.064 nN/nm and the frequency of oscillation was 1 Hz.

Conclusions

PEO of high molecular weight was physically adsorbed to AFM cantilever probes. These probes initially produced steric exclusion forces in 0.1 M KNO_3 upon compression of the adsorbed PEO chains by the sample. The observation of a steric exclusion force commencing at a distance less than 2.5 - 3.5 R_g was an indication of expulsion of the PEO from the contact area. Also, observed was the presence of a long range attractive force upon retraction of the sample. The origin of this force is likely due to bridging that occurs between the uncoated surface and the physically adsorbed PEO. PEO physically adsorbed to the sample also demonstrated steric exclusion forces, expulsion of PEO from the contact area and bridging of the PEO to the uncoated probe.

Force-distance curves obtained in 0.1 M KNO_3 containing 0.1 % PEO showed no repulsive steric exclusion force initially. After 8 hr of incubation, a steric expulsion force was observed which commenced at a distance near 6 R_g. Often curves which demonstrated bridging or entanglement were visible.

PEG with a molecular weight of 2 kDa was chemically bound to the surface of the silicon nitride tip, as shown by ESCA. Magnitude calculations from GOLIAD indicate that a steric repulsive force of 6 nN would be present at a separation distance of 0.6 nm. The force-distance plots do not show a monotonically increasing repulsive force, indicating that there is much less than one equivalent monolayer of PEG on the surface or the force gradient is larger than the cantilever spring constant at a measurable force.

Acknowledgements

The authors wish to thank Y.-S. Lin for assistance in the PEG surface binding chemistry, A. Pungor and E. W. Stroup for technical assistance and helpful discussions, and P. Barnevald and late J. M. H. M. Scheutjens for providing us with the computer modelling programs POLAD and GOLIAD. We also are grateful for financial support provided by the NIH (Grant HL-44538-02) and the University of Utah Graduate Research Committee.

Literature Cited

1. Binnig, G.; Quate, C. F.; Gerber, C. *Phys. Rev. Lett.* **1986** *56*, 930-933.
2. Martin, Y.; Williams, C. C.; Wickramasinghe, H. K. *J. Appl. Phys.* **1987** *61*, 4723-4729.
3. Weisenhorn, A. L.; Hansma, P. K.; Albrecht, T. R.; Quate, C. F. *Appl. Phys. Lett.* **1989** *54*, 2651-2653.
4. Albrecht, T. R. *Ph.D. Dissertation,* Stanford University 1989.
5. Pashley, R. M. *J. Colloid Interface Sci.* **1981** *80*, 153-162.
6. Hartmann, U. Ultramicroscopy **1992** *42-44*, 59-65.
7. McClelland, G. M.; Erlandsson, R.; Chiang, S. In *Review of Progress in Quantitative Nondestructive Evaluation,* **1987** *6B*, 1307-1314.
8. Stern, J. E.; Terris, B. D.; Mamin, H. J.; Rugar, D. *Appl. Phys. Lett.* **1988** *53*, 2717-2719.
9. Ducker, W. A.; Senden, T. J.; Pashley, R. M. *Nature* **1991** *353*, 239-241.
10. Martin, Y.; Wickramasinghe, H. K. *Appl. Phys. Lett.* **1987** *50*, 1455-1457.
11. Grütter, P.; Meyer, E.; Heinzelmann, H.; Rosenthaler, L.; Hidber, H.-R.; Güntherodt, H.-J. *J. Vac. Sci. Technol.* **1988** *A6*, 279-282.
12. Klein, J.; Luckham, P. *Nature* **1982** *300*, 429-431.
13. Claesson, P. M.; Gölander, C.-G. *J. Colloid Interface Sci.* **1987** *117*, 366-374.
14. Taunton, H. J.; Toprakcioglu, C.; Fetters, L. J.; Klein, J. *Nature* **1988** *332*, 712-714.

15. de Gennes, P. G. *Macromolecules* **1981** *14*, 1637-1644.
16. Milner, S. T. *Science* **1991** *251*, 905-914.
17. Lea, A. S.; Pungor, A.; Hlady, V.; Andrade, J. D.; Herron, J. N.; Voss Jr., E. W. *Langmuir* **1992** *8*, 68-73.
18. Marchant, R. E.; Lea, A. S.; Andrade, J. D.; Bockenstedt, P. *J. Colloid Interface Sci.* **1992** *148*, 261-272.
19. Harris, M. J.; Struck, E. C.; Case, M. G.; Paley, M. S.; Van Alstine, J. M.; Brooks, D. E. *J. Polymer Sci., Polym.Chem.Ed.* **1984** 22, 341-352.
20. Lin, Y.-S. and Hlady, V. *Book of Abstracts, Part I,* 203rd ACS National Meeting, ACS, San Francisco, 1992, abst. # COLL361.
21. Scheutjens, J. M. H. M.; Fleer, G. J.; Cohen Stuart, M. A. *Colloids Surfaces* **1986** *21*, 285-306.
22. Milner, S. T. *Europhys. Lett.* **1988** *7*, 695-699.
23. Barnevald, P. A.; *Doctoral Thesis*, Agricultural University, Wagengingen, The Netherlands, 1991.
24. The ESCA data were actually obtained from the chip holding the cantilever. Being of the same surface chemistry, the tip and the chip are expected to produce similar results.
25. Sobolewski, M. A.; Helms, C. R. *J. Vac. Sci. Technol.* **1988** *A6*, 1358-1362.
26. Bergström, L.; Bostedt, E. *Colloids Surfaces* **1990** *49*, 183-197.
27. Bergström, L.; Pugh, R. J. *J. Am. Ceram. Soc.* **1989** *72*, 103-109.
28. Jaffrezic-Renault, N.; De, A.; Clechet, P.; Maaref, A. *Colloids Surfaces* **1989** *36*, 59-68.
29. Fleer, G. J.; Lyklema, J. In *Adsorption from Solution at the Solid/Liquid Interface;* Parfitt, G. D. and Rochester, C. H., Eds., Academic Press: London, UK, 1983, pp 153-220.
30. Ben Ouada, H.; Hommel, H.; Legrand, A. P.; Balard, H.; Papirer, E. *J. Colloid Interface Sci.* **1988** *122*, 441-449.
31. Cohen Stuart, M. A.; Waajen, F. H. W. H.; Cosgrove, T.; Vincent, B.; Crowley, T. L. *Macromolecules* **1984** *17*, 1825-1830.
32. Albrecht, T. R. *Ph.D. Thesis,* Stanford University, 1991.
33. Drake, B.; Prater, C. B.; Weisenhorn, A. L.; Gould, S. A. C.; Albrecht, T. R.; Quate, C. F.; Cannell, D. S.; Hansma, H. G.; Hansma, P. K. *Science* **1989** *243*, 586-1589.
34. Klein, J.; Luckham, P. F. *Nature* **1984** *308*, 836-837.

RECEIVED February 23, 1993

INDEXES

Author Index

Affiliation Index

Subject Index

Production: Meg Marshall
Indexing: Deborah H. Steiner
Acquisition: Anne Wilson
Cover design: Alan Kahan

Printed and bound by Maple Press, York, PA

Bestsellers from ACS Books

The ACS Style Guide: A Manual for Authors and Editors
Edited by Janet S. Dodd
264 pp; clothbound ISBN 0–8412–0917–0; paperback ISBN 0–8412–0943–X

The Basics of Technical Communicating
By B. Edward Cain
ACS Professional Reference Book; 198 pp;
clothbound ISBN 0–8412–1451–4; paperback ISBN 0–8412–1452–2

Chemical Activities (student and teacher editions)
By Christie L. Borgford and Lee R. Summerlin
330 pp; spiralbound ISBN 0–8412–1417–4; teacher ed. ISBN 0–8412–1416–6

Chemical Demonstrations: A Sourcebook for Teachers,
Volumes 1 and 2, Second Edition
Volume 1 by Lee R. Summerlin and James L. Ealy, Jr.;
Vol. 1, 198 pp; spiralbound ISBN 0–8412–1481–6;
Volume 2 by Lee R. Summerlin, Christie L. Borgford, and Julie B. Ealy
Vol. 2, 234 pp; spiralbound ISBN 0–8412–1535–9

Chemistry and Crime: From Sherlock Holmes to Today's Courtroom
Edited by Samuel M. Gerber
135 pp; clothbound ISBN 0–8412–0784–4; paperback ISBN 0–8412–0785–2

Writing the Laboratory Notebook
By Howard M. Kanare
145 pp; clothbound ISBN 0–8412–0906–5; paperback ISBN 0–8412–0933–2

Developing a Chemical Hygiene Plan
By Jay A. Young, Warren K. Kingsley, and George H. Wahl, Jr.
paperback ISBN 0–8412–1876–5

Introduction to Microwave Sample Preparation: Theory and Practice
Edited by H. M. Kingston and Lois B. Jassie
263 pp; clothbound ISBN 0–8412–1450–6

Principles of Environmental Sampling
Edited by Lawrence H. Keith
ACS Professional Reference Book; 458 pp;
clothbound ISBN 0–8412–1173–6; paperback ISBN 0–8412–1437–9

Biotechnology and Materials Science: Chemistry for the Future
Edited by Mary L. Good (Jacqueline K. Barton, Associate Editor)
135 pp; clothbound ISBN 0–8412–1472–7; paperback ISBN 0–8412–1473–5

For further information and a free catalog of ACS books, contact:
American Chemical Society
Distribution Office, Department 225
1155 16th Street, NW, Washington, DC 20036
Telephone 800–227–5558